Praise for
Our Fragile Mo

"Deeply researched, sprawling in scope, and with insights and surprises on every page. This is the sort of historical understanding that leads to wisdom."

—Seth Godin, founding editor of *The Carbon Almanac*

"This detailed and yet marvelously readable look at our climatic past offers us the information we need to understand our climatic future—and more importantly, to act to shape that future in the here and now." —Bill McKibben, author of *The End of Nature*

"Michael Mann has a tremendous depth of knowledge about the history of our planet's climate, which is why his words of warning and optimism are so important. This book provides important lessons from humanity's past to empower readers to help protect our future." —Al Gore, former US vice president

"A gripping tale of Earth's climate history, this book is a must-read for every global citizen. It dispels common climate myths with surgical clarity and provides an essential roadmap to understanding our past and choosing our future."

—Katharine Hayhoe, climate scientist,
distinguished professor at Texas Tech University,
UN Champion of the Earth, author of *Saving Us*

"Reading *Our Fragile Moment* is like taking a spectacular hike through billions of years of Earth's climate history with one of the great scientists of our time. Oh look—there's the meteor that wiped out the dinosaurs! There's the great ocean conveyor! There's the Rossby waves! When you reach the summit of Mann's wonderful book, you will understand just how rare and beautiful our moment is—and why we need to fight harder to protect it."

—Jeff Goodell, author of *The Heat Will Kill You First*

"Mann shows over the last few hundreds of millions of years, Earth has been snow-ball cold, tropic hot, rainforest wet, and desert dry. Its atmosphere has been oxygen poor, oxygen rich, or choked with deadly gas. But Earth has never been through anything quite like humankind. Our current comfortable climate is disappearing—because of us. It's cause for thundering alarm but is not cause for despair or doomist gloom. It's time for action. Don't believe me? Read this book."

—Bill Nye, science educator and CEO, the Planetary Society

"Mann has masterfully woven the climate story from our past to the future. Drawing upon a wealth of data, research, and expertise, he slays the persistent zombie theories that climate scientists ignore historical context."

—Dr. Marshall Shepherd, international expert in weather and climate and distinguished professor of geography and atmospheric sciences, University of Georgia

"Mann knows climate science as well as anyone on Earth, and knows how to write with clarity, brevity, and wit. *Our Fragile Moment* is a book which will benefit all readers from novices to informed professionals. Mann presents a riveting, readable, and instructive narrative of Earth's climate changes over timescales ranging from geologic to human/historical and artfully explains what lessons we can draw to help us navigate this new epoch of human-altered climate. He shows how an honest, informed look at planetary history serves as both a defense against doomism and a call to action to forge a livable world that is still well within our grasp."

—David Grinspoon, astrobiologist and author of *Earth in Human Hands*

Our
Fragile
Moment

Our
Fragile
Moment

How Lessons from
Earth's Past Can
Help Us Survive the
Climate Crisis

Michael E.
Mann

PUBLICAFFAIRS
New York

PublicAffairs
Hachette Book Group
1290 Avenue of the Americas, New York, NY 10104
www.publicaffairsbooks.com
@Public_Affairs

Printed in the United States of America

First Edition: September 2023
First Trade Paperback Edition: October 2024

Published by PublicAffairs, a subsidiary of Hachette Book Group, Inc. The PublicAffairs name and logo is a registered trademark of the Hachette Book Group.

The Hachette Speakers Bureau provides a wide range of authors for speaking events. To find out more, go to hachettespeakersbureau.com or email HachetteSpeakers@hbgusa.com.

PublicAffairs books may be purchased in bulk for business, educational, or promotional use. For more information, please contact your local bookseller or the Hachette Book Group Special Markets Department at special.markets@hbgusa.com.

The publisher is not responsible for websites (or their content) that are not owned by the publisher.

Print book interior design by Jeff Williams

The Library of Congress has cataloged the hardcover edition as follows:
Names: Mann, Michael E., 1965– author.
Title: Our fragile moment : how lessons from Earth's past can help us survive the climate crisis / Michael E. Mann.
Description: First edition. | New York : PublicAffairs, 2023. | Includes bibliographical references and index.
Identifiers: LCCN 2023016363 | ISBN 9781541702899 (hardcover) | ISBN 9781541702912 (ebook)
Subjects: LCSH: Climatic changes—History. | Climate change mitigation.
Classification: LCC QC903 .M36255 2023 | DDC 363.738/746—dc23/eng20230706
LC record available at https://lccn.loc.gov/2023016363

ISBNs: 9781541702899 (hardcover), 9781541702912 (ebook), 9781541702905 (paperback)

LSC-C

Printing 1, 2024

Michael Mann dedicates this book to

his wife, Lorraine Santy, and daughter,
Megan Dorothy Mann,

and to the memory of his brother,
Jonathan Clifford Mann,

and mother, Paula Finesod Mann

Contents

Acronyms

AIS	Arctic Ice Sheet
AMO	Atlantic Meridional Overturning
AMOC	Atlantic Meridional Overturning Circulation
AO	Arctic Oscillation
AOC	Atlantic Overturning Circulation
BAU	business as usual
BCE	Before Common Era
BP	Before Present
C, C-12, C-13	carbon, carbon with 6 protons-6 neutrons, carbon with 6 protons-7 neutrons
C	Celsius (after astronomer Anders Celsius)
CE	Common Era
CFC	Chloro-Fluoro-Carbon
CH_4	methane, or natural gas
CIS	Cenozoic Ice Age
CNN	Cable News Network
CO_2	carbon dioxide
COP	Conference of the Parties
COVID-19	Coronavirus disease of 2019
DDT	Dichloro-Diphenyl-Trichloroethane
DNA	Deoxyribo-Nucleic Acid
EAIS	East Antarctic Ice Sheet
ECS	Equilibrium Climate Sensitivity
ENSO	El Niño Southern Oscillation
EO	Eocene-Oligocene
ESS	Earth System Sensitivity
ETS	Emissions Trading Scheme
F	Fahrenheit (after physicist Daniel Fahrenheit)

FYS	Faint Young Sun
GBR	Great Barrier Reef
GHG	greenhouse gas
GIS	Greenland Ice Sheet
GISS	Goddard Institute for Space Studies
GLH	Glacial Lake Hitchcock
GMI	George Marshall Institute
GOE	Great Oxidation Event
GRL	Geophysical Research Letters
GtC	gigatons of carbon
HBO	Home Box Office
HE (HHE)	Hothouse Earth
HG	Huronian Glaciation
INF	Intermediate-Range Nuclear Forces
IPCC	Intergovernmental Panel on Climate Change
ISIS	Islamic State of Iraq and Syria
K	Kelvins (Celsius-sized degrees above absolute zero)
K-Pg	[K]retaceous-Paleogene
K-T	[K]retaceous-Triassic
LALIA	Late Antique Little Ice Age
LDEO	Lamont Doherty Earth Observatory
LGM	Last Glacial Maximum
LIA	Little Ice Age
LIC	Laurentide Ice Sheet
LOA	Library of Alexandria
LPG	Late Proterozoic Glaciation
MAR	Mid-Atlantic Ridge
MICI	Marine Ice Cliff Instability
MISI	Marine Ice Sheet Instability
MIT	Massachusetts Institute of Technology
MP	Member of Parliament
MP	Montréal Protocol
MPT	Mid-Pleistocene Transition
NAIP	North Atlantic Igneous Province
NAO	North Atlantic Oscillation
NAS	National Academy of Sciences
NASA	National Aeronautics and Space Administration

NCAR	National Center for Atmospheric Research
NORAD	North American Air Defense command
NPR	National Public Radio
NSE	Neoproterozoic Snowball Earth
OKE	Old Kingdom of Egypt
O-16, O-18	oxygen with 8 protons-8 neutrons, oxygen with 8 protons-10 neutrons
PETM	Paleo-Eocene Thermal Maximum
pH	power (exponent) of hydrogen (*pouvoir hydrogen*)
Ph.D.	Doctor, of Philosophy (or Piled Higher and Deeper)
PICIR	Potsdam Institute for Climate Impact Research
ppm	parts per million
PSE	Paleoproterozoic Snowball Earth
P-T	Permian-Triassic
RCP	Representative Concentration Pathways
SBE	Snowball Earth
SDI	Strategic Defense Initiative ("Star Wars" weapons)
SEPP	Science and Environmental Policy Project
SL	sea level
SASM	South Asian Summer Monsoon
SIO	Scripps Institute for Oceanography
SPM	Summary for Policy Makers
SST	Sea Surface Temperature
SXSW	South by Southwest
SUV	sport utility vehicle
Texas A&M	Texas Agricultural and Mechanical University
TTAPS and TTAPS2	Turco, Toon, Ackerman, Pollack, and Sagan papers
UAE	United Arab Emirates
UC Berkeley	University of California, Berkeley
UMass	University of Massachusetts
UNHCR	United Nations High Commissioner on Refugees
UNSW	University of New South Wales
UV	Ultra Violet
UVa	University of Virginia
WAIS	West Antarctic Ice Sheet
YD	Younger Dryas

Introduction

We live on a Goldilocks planet. It has water, an oxygen-rich atmosphere, and an ozone layer that protects life from damaging ultraviolet rays. It is neither too cold nor too hot, seemingly just right for life. Despite our ongoing search—which, with the recent advent of the James Webb telescope, now extends out nearly fourteen billion light years—we have thus far found no other planet in the universe with such benevolent conditions. It's almost as if *this* planet, Earth, was custom made for us. And yet it wasn't.

For the vast majority of its 4.54 billion years, Earth has proven it can manage just fine without human beings. The first hominids—proto-humans—emerged a little more than two million years ago. Only during the past 200,000 years have modern humans walked the Earth. And human civilizations have existed for only about 6000 or so years, 0.0001 percent of Earth's history—a fleeting moment in geological time.

What is it that made this fragile yet benevolent moment of ours possible? Ironically, it's the very same thing that now threatens us: climate change. The asteroid impact sixty-five million years ago that generated a global dust storm chilled the planet, killing off the dinosaurs and paving the way for our ancestors, tiny shrew-sized proto-mammals that scurried about, hiding from their saurian predators. With the dinosaurs no longer around, these critters could now

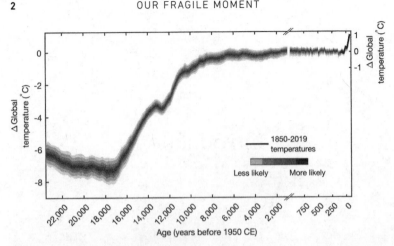

Figure 1. Estimated global temperature changes over the past 24,000 years. The fragile moment is defined by the period from around 6000 years ago to the mid-twentieth century (the "zero" of the time axis).

come out from the shadows, fill new niches, and gradually branch out to produce primates, apes, and eventually us. Though such an event would prove devastating for modern human civilization if it happened today, our real and urgent threat is from fossil fuel burning and carbon pollution, and it is warming, not cooling.

Climate has shaped and guided us from the start. The drying of the tropics as the planet cooled during the Pleistocene epoch of the past 2.5 million years created a niche for early hominids, who could hunt prey as forests gave way to savannas in the African tropics. Yet drying today threatens drought and wildfire in many regions. The sudden cooling episode in the North Atlantic Ocean 13,000 years ago known as the Younger Dryas, which occurred just as Earth was thawing out of the last ice age, challenged hunter-gatherers, spurring the development of agriculture in the Fertile Crescent. A similar North Atlantic cooling event looms today as Greenland ice melts, freshens the waters of the North Atlantic, and disrupts the northward ocean conveyor current system. It could threaten fish populations and impair our ability to feed a hungry planet. The Little Ice Age of the sixteenth to nineteenth centuries led to famines and pestilence for much of Europe, and contributed to the collapse of the Norse colonies. Yet it was a boon for some, such as the Dutch, who were able to take advantage of stronger winds to shorten their ocean voyages. The Dutch West and East India Companies became the dominant maritime trading companies, holding a near monopoly on European shipping routes to South and North America, Africa, Australia, and New Zealand. They seemingly ruled the world. For a while. Just like the dinosaurs did. For a while.

As we can see, the story of human life on Earth is a complicated one. Climate variability has at times created new niches that humans or their ancestors could potentially exploit, and challenges that caused devastation, then spurred innovation. But the conditions that allowed humans to live on this Earth are incredibly fragile, and there's a relatively narrow envelope of climate variability within which human civilization remains viable. Today, our massive societal infrastructure supports more than eight billion people, an order of magnitude beyond the natural "carrying capacity" of our planet (the resource limit of what our planet can provide in the absence of human technology). The resilience of this infrastructure depends on conditions remaining the same as those that prevailed during its development.

The concentration of carbon dioxide (CO_2) in the atmosphere today is the highest since early hominids first hunted on the African savannas. It is now already outside the range during which our civilization arose. If we continue to burn fossil fuels, it is likely that the planet will warm beyond the limit of our collective adaptive capacity. How close are we to the edge? In the pages that follow, I set out to answer that question.

We'll look at how we have arrived here, and the incredible gift of a stable climate that the planet gave us along the way so that we, humans, could not just exist but thrive. And we'll learn how our civilization will be imperiled if we continue on our current path. We'll delve into the field known as *paleoclimatology*—the study of prehistoric climates—which offers crucial lessons as we contend with the greatest challenge we've faced as a species. You already, no doubt, know that we face a climate crisis. In the following pages, I'll arm you with the knowledge necessary to fully appreciate the extent of the unfolding threat, while emboldening you to act before it truly does become too late. Only by understanding the climate changes of the past and what they tell us about the circumstances that allowed us to thrive, can we appreciate two seemingly contradictory realities. On the one hand, there is the absolute fragility of this moment in time—driven home on a daily basis by each devastating wildfire and every "once in a century" hurricane or 110°F day, collective signs that we seem to be slipping into the chasm of an unlivable planet. On the other hand, however, the study of Earth's history betrays some degree of climate resilience. Climate change is a crisis, but a *solvable* crisis.[1]

An important point we'll come back to often throughout this book is this: we must embrace *scientific uncertainty*. The scientific process builds on itself. New data come to light that help us refine our understanding. Sometimes it changes our previous understanding. Contrarians insist that this uncertainty is a reason for climate inaction, the implication being that we can't trust it, or we might somehow overreact in a way that, for example, could hurt the economy. But just the opposite is true. Many key climate impacts—the increase in deadly and devastating extreme weather events, the loss of glacial ice, and the resulting inundation of our coasts—have already exceeded the earlier scientific projections. Uncertainty isn't our friend. It is, however, a very good reason for even greater precaution and more concerted action.

A consequence of this uncertainty, as we'll see, is that the answers aren't always cut and dry—this is particularly true as we go back in time and the data become both sparser and fuzzier. Our instinct is to try to come up with simple analogs and definitive conclusions. But science doesn't work that way, and a complex system like Earth's climate certainly doesn't work like that. So we must embrace nuance, too—and indeed it is one of our greatest tools as we seek answers to the key questions about our climate past and our climate future.

Different scientific studies often come to at least modestly different conclusions. It is only by assessing the collective evidence across numerous scientific studies that we reach more firm conclusions and begin to establish scientific consensus. I've always loved this story told by Ira Flatow, the amiable host of NPR's *Science Friday*, about a fact-finding congressional inquiry into the potential threat posed by supersonic air travel during the early 1970s:

Senator Edmund Muskie (D-ME) was the chairman of the committee assigned to find the answers to these questions. He, in turn, appointed an august committee from the National Academy of Sciences (NAS) to study the issue. Six months later they were to report to the congressional committee. All the newspapers were there and the cameras were rolling.

The committee's chief scientist said, "Senator, we're ready to testify," and Muskie responded, "Okay, tell me what the answer is. Is this going to be a danger?" The scientist then slapped down his giant sheaf of papers on the desk and said, "I've got these papers here that definitely tell us this is going to be a danger." Muskie was ready to conclude right there, but then the NAS scientist interjected, "On the other hand, I have another set of papers over here that says these papers aren't good enough to know the answer." In exhaustion, the senator looked up and yelled, "Will somebody find me a one-handed scientist?!"[2]

It's a cute story, but with a serious lesson. Everybody wants a "one-handed scientist," but that's not how science works.

Complicating matters further is that press releases and media coverage tend to emphasize "blockbuster" studies: ostensibly shocking new discoveries that garner clicks and pageviews. So we get the so-called whiplash effect, where we're told one week about a study, for example, that shows that eating chocolate or drinking coffee or wine (basically all the good stuff life offers) is healthy, only to read a headline the following week about a new study insisting it's bad for you.[3]

As a result, we get a skewed view of scientific understanding as more polarized and more mercurial than it actually is. The phenomenon is readily seen in the climate discourse, where we're told one week, for example, that the Greenland Ice Sheet—and all the sea level rise that comes with it—may be on the verge of collapse, while a study the following week suggests it's more stable than we thought. We're frequently bombarded with dire headlines about "doomsday glaciers" and "methane bombs" that belie the still dire but more nuanced and, importantly, *far less hopeless* picture that emerges from an objective assessment of the underlying scientific evidence.

Keeping uncertainty and all its implications in mind, we'll look at the big question on everybody's mind: Are we doomed? The answer, as we'll learn, is that it is entirely up to us. The collective evidence from the paleoclimate record—the record of Earth's past climatic changes—actually provides a blueprint for what we need to do to preserve our fragile moment. The greatest threat to meaningful climate action today is no longer denial, but despair and doomism, premised on the flawed notion that it is too late to do anything. Our review of the paleoclimatic record will tell us otherwise.

There is a duality that governs the human species and the climate it enjoys. Human actions, particularly the burning of fossil fuels and the generation of carbon pollution, have impacted the trajectory that our climate has taken over the past two centuries, but the longer-term trajectory of our climate has also impacted *us*. It's what got us here. By looking back at that trajectory, we can gain insights into what futures are possible. In my previous book *The New Climate War*, I examined the lobbying efforts over the past half century by fossil fuel companies and their enablers that have prevented us from thus far taking the actions necessary to avert catastrophic climate change. Thanks to the efforts of those corporations, we're now coming up against the boundary of habitable life for us humans.

In this book, I'm inverting this perspective. We're going to look at the influence that Earth's climate history has had on us and what we can learn from it. But keep in mind that paleoclimate is only one line of evidence. It will not and cannot address all of the questions we might have about human-caused climate change, if for no other reason than there is no perfect analog in our past for what we potentially face in the future. But together with insights from the modern climate record and guidance from state-of-the-art models of Earth's climate system, it informs our assessment of just how tenuous this moment is, underscoring both the urgency of actions to mitigate, and adapt to, the heightening climate crisis we face and the agency that we still possess in averting disaster.

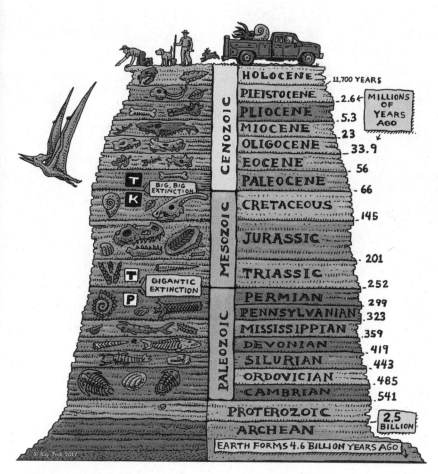

Figure 2. The geological timescale.

1

Our Moment Begins

We are at a crossroads in human history. Never before
has there been a moment so simultaneously perilous and
promising. We are the first species to have taken evolution into our own hands.

—CARL SAGAN, *Broca's Brain*

Imperiled today by planetary warming of our own making, we find
no small amount of irony in our current plight. For we, in fact, owe
our very existence to climate change, albeit the *natural* sort. A violent
asteroid impact here, a sudden warming spike there, with a collision
of tectonic plates and a collapsing ocean conveyor thrown into the
mix. We wouldn't have arrived at this moment without a remarkable
series of climate episodes and accidents.

We Emerge

Climate change has shaped us from the beginning. It's a simple statement, but the story is not. Not all evolutionary or societal trends
are driven by environmental changes, let alone climate change. Some
simply reflect the slow but steady progress of natural selection acting
on environmental conditions, chance discoveries, and innovations.
But there is no doubt that some of the key developments that made

us what we are today were driven by climate events. Let's start at the beginning.

One could argue that the *possibility* of human life began when the first living organism emerged from the primordial ooze somewhere around four billion years ago. But really, it began in earnest several billion years later—sixty-six million years ago to be more precise—when a giant asteroid struck Earth. It was almost eight miles wide, traveling 30,000 miles per hour (more than thirty times faster than the speed of sound), creating a hundred-mile-wide crater that lies beneath the waters off the coast of the Yucatan Peninsula. This epic collision ejected a massive cloud of debris into the atmosphere, blocking out sunlight and rapidly cooling the planet. Roughly eighty percent of all animal species including the non-avian dinosaurs (that "non-avian" qualifier is required because birds, as a direct lineage of the dinosaurs, are in fact a surviving subclass) disappeared.

As we'll see throughout this story, tragedy for some often meant opportunity for others. A key consequence of this catastrophic event was that it eliminated the main predators of our extreme distant ancestors, the small rodent-like mammals that had scurried about hiding among the rocks. They could come out of hiding and occupy new niches. The collision marked the end of the Mesozoic era—the age of dinosaurs—and the beginning of the Cenozoic era—the *age of mammals*.

Roughly fifty-six million years ago, just a brief (in geological time) ten million years after the demise of the dinosaurs, climate change— that is, *naturally occurring* climate change—generated yet another challenge for life. In this case, it was a sudden warming spike known as the Paleocene-Eocene Thermal Maximum, or PETM, which occurred early in the twenty-million-year-long Eocene epoch. It created evolutionary pressures that opened a niche for a whole new order of mammals: the *primates*. The first one was a primitive lemur-like creature named *Dryomomys*. She was indeed our distant relative, though at five inches long and on a fruit diet, she would be awkward to invite to Thanksgiving dinner.[1]

The warm, humid greenhouse climates of the early and mid-Eocene also favored an increase in plant diversity, which in turn created new environmental niches for primates. Wet tropical and subtropical forests allowed the arboreal early primates to diversify and disperse across North America, Eurasia, and Africa. The climate then slowly cooled over the course of the mid- and late Eocene and into the Oligocene epoch, which began roughly thirty-four million years ago.

What drove this cooling trend? Back in the early 1990s, leading paleoclimate scientists Maureen Raymo and William Ruddiman argued that it was the collision of the Indian and Asian tectonic plates, beginning in the early Eocene. That collision pushed up and created the Tibetan plateau, also building the towering Himalayan mountain range (and the majestic Mt. Everest). The warm, moisture-laden air coming off the Indian ocean collides with the mountain range and rises upward toward the peaks, causing the moisture to condense into rainfall as it rises and cools in the atmosphere. Today we know this system as the South Asian summer monsoon.[2]

More rainfall means more weathering of rocks; CO_2 from the atmosphere dissolves in streams and rivers, where it turns into carbonic acid and dissolves rocks. Silicate rocks known as feldspars, for example, dissolve into clay and calcium and carbonate ions (very small charged molecules). These materials run off into streams and rivers and eventually into the ocean. All this carbon drawn down from the atmosphere and buried in the ocean weakens the greenhouse effect and cools down the planet.

It was during the early Oligocene that the first hints of something we might call an *icehouse* climate began to emerge. Ice sheets formed, first in Antarctica around thirty-four million years ago and later in North America and Greenland.[3]

The CO_2 drawdown and cooling continued on into the subsequent Miocene epoch that began twenty-four million years ago. The cooler conditions led to the retreat of subtropical forests, which were replaced by woodlands—more open environments that consist of a mix of trees, grasses, shrubs, and other plants, creating a niche for primates that spent more time on the ground and less time swinging from trees. Welcome to the "planet of the apes." Orangutans appeared during the mid-Miocene. Gorillas appeared a few million years later, and the first chimps a few million years after that. With

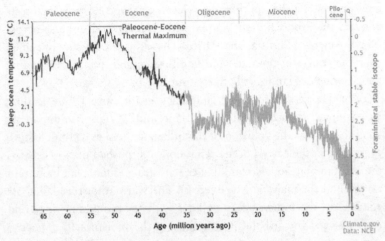

Figure 3. Estimated global temperature changes over the past sixty-five million years.

primates that are now partly upright, use primitive tools, and have a more complex social organization—the great apes, or *hominids*—we are inching closer and closer to our species, *Homo sapiens.*

As forests and woodlands disappeared in Eurasia from six to twelve million years ago, so did the populations of great apes. There was an exodus of hominids to southeast Asia and to Africa, where some eventually evolve into *hominins*—an even more select group that includes modern humans, extinct human species, and our immediate ancestors. Meanwhile, the carbon drawdown, and cooling, continued. Grasslands expanded. So did ice sheets, which began to form in the Northern Hemisphere, in Greenland and North America.

At five million years ago we had entered into the Pliocene. This was the last time the level of greenhouse gases in the atmosphere was comparable to today, between 380 and 420 parts per million (ppm) in the atmosphere. Yet somehow, the planet was actually 3.5–5°F warmer than today, and the sea level could have been as much as thirty feet higher. What gives? We'll explore this seeming paradox in a bit.[4]

Homo sapiens weren't yet on the scene, but our direct ancestors were. The first upright walking hominins, like Ardipithecus (and, soon, Australopithecus), evolved from earlier hominids on the African continent. Cooling continued, and expansive subtropical savannas and grasslands replaced forests and woodlands. Though these environments were not well suited to the great apes, who retreated to wetter tropical locales (where they remain to this day), they were ideal for the newly evolved hominins, who carved out a niche by walking upright. They were omnivores who supplemented the fruit and nuts favored by their ape ancestors with edible grasses and sedges and the meat they obtained by hunting large game animals in packs on the savannas.

Fast-forward to 2.6 million years ago and we are headed directly into the icehouse—the Pleistocene epoch. Cooling continued and ice sheets began to take hold now in the Northern Hemisphere. Several species of our own genus *Homo* now roamed the African savannas. Some evolved into faster runners and more effective hunters, taking advantage of the expanded grasslands. Some used rudimentary stone tools. Some developed bigger, better brains and would ultimately evolve into our very own species, *Homo sapiens*.

By 700,000 years before the present, the climate had cooled yet further, and Earth experienced more extensive glaciation with ice sheets extending well down into North America and Eurasia. These expanded ice sheets displaced the jet stream toward the equator, cooling and drying the subtropics and tropics, including large parts of Africa where the hominins resided. Disrupted climate patterns may have naturally selected for species with higher-powered brains that could develop strategies to contend with the severe challenges created by the changing climate, including the design of more-sophisticated stone tools and development of increasingly more complex social communities, like hunting groups using better-designed spears and sophisticated group hunting strategies to more efficiently hunt game when other food sources were scarce.[5]

Larger ice sheets fundamentally changed the dynamics of the climate system itself, generating larger, slower swings between cold *glacial* periods (ice ages) with extensive ice sheets and warmer *interglacial* periods with greatly diminished ice. These swings are tied to astronomical cycles governing Earth's orbit relative to the Sun, especially the roughly 100,000-year cycle's *eccentricity* (how circular versus elliptical the Earth's annual orbit around the Sun is). But their magnitude is determined by how cold and icy the planet can get.[6]

The long-term cooling trend proceeded from around 700,000 years ago, continuing for the next several hundred thousand years. That lead to the intermittent growth of larger and larger ice sheets, causing increasing climate disturbance in the African tropics and subtropics during the ensuing glacial/interglacial cycles. The most recent complete cycle was the largest swing of all, ranging from the extreme warmth of the Eemian period starting 130,000 years ago (which at its peak likely exceeded even today in warmth) to the bitter cold of the Last Glacial Maximum roughly 21,000 years ago when an ice sheet covered what is now New York City. The global temperature change during that swing was about 9°F, and twice that over middle and high latitudes, owing to the amplifying effects of growing or shrinking ice—it's a so-called positive feedback loop—something that we'll see is critical in climate change. These huge swings in climate put even greater selective pressure on bigger and better brains that could devise ever more clever coping mechanisms to deal with the challenges presented by climate extremes.

And so, our moment finally arrives. Bones of primitive *Homo sapiens* first appeared 300,000 years ago in Africa, with skulls suggesting brains of similar size to our own. Anatomically modern *Homo sapiens* appeared 200,000 years ago, and skulls from 100,000 years ago suggest brains that were indistinguishable from our own in all respects, including both size and shape. Between 100,000 and 200,000 years ago, *Homo sapiens* were collecting and cooking shellfish and making fishing tools. They were using language. They had become us. But though modern humans had finally arrived, with the help of an asteroid impact, a long-term cooling trend, and the huge seesaw cycles of warming and cooling that marked the late Pleistocene, it would take another series of climate events and accidents to yield the innovation essential for the emergence of human *civilization*.[7]

In the Wilderness

For our first hundred thousand years we were out in the wilderness— literally. Early *Homo sapiens* existed as nomadic, hunter-gatherer tribes. They grew in numbers and expanded into other continents including Europe and Asia, following the migrations of other archaic *Homo* subspecies, the Neanderthals and Denisovans, into those regions. There is evidence that we both fought with and at times interbred with these other subspecies (their genes live on in many of us today). But mostly, we outcompeted them, with some help from climate change. Some archeologists believe that Neanderthals were ill equipped to withstand an extended cold period in Europe around 40,000 years ago because they were so reliant on the hunting of particular large game animals whose populations were diminished by the effects of cooling.[8]

In the ensuing 40,000 years we *Homo sapiens* would continue to use our big brains to develop increasingly sophisticated stone tools, don clothing, and communicate with each other through complex language. We engaged in ceremonial burying of the dead, adorned ourselves with ornamental necklaces, and crafted figurines that celebrated the human form. On cave walls we painted murals that illustrated our ways of living and, especially, the animals we hunted for food. We developed seafaring and migrated to distant lands like Australasia and, eventually, the Americas.[9]

We had come a long way. But we were not yet *civilized*. Constrained by our nomadic lifestyle, we lacked the ability to establish permanent or semi-permanent settlements, complex social hierarchies, or the division of labor necessary to support large, sustainable populations. That would all change as the result of a climate accident that occurred during the great meltdown that was the end of the last ice age.

There's a sort of "glitch" in the climate system that can occur when a large amount of freshwater is suddenly released into the oceans. The trigger in this case was the rapid melting of the expansive Laurentide Ice Sheet, a continental-sized glacier that covered the upper half of North America, its southern edge reaching as far south as Chicago and New York City. The meltdown began about 15,000 years ago as the exit from the last ice age was accelerating. It would continue for about two millennia. This warm interval is known as the Bølling-Allerød interstadial.

The rapid ice melt led to a series of massive freshwater pulses exiting to the North Atlantic Ocean. Some have speculated that the flood myths found among various cultures, including the Noachian Deluge in the book of Genesis and the great flood described in the ancient Mesopotamian epic of Gilgamesh, could have their origins in one of these meltwater events. One particularly sharp pulse, creatively titled "meltwater pulse 1A," occurred between 13,500 and 14,700 years ago, and it dumped enough water into the ocean to raise global sea levels by a whopping forty-four feet in just 300 years.[10]

Some of the water from 1A, however, pooled up in a huge glacial lake, known as Lake Agassiz, that formed at the southern end of the Laurentide in south central Canada. A massive ice dam prevented the water from escaping to the ocean. Suddenly 12,900 years ago, the dam appears to have broken. The massive release of glacial meltwater would have dramatically freshened the surface of the northern region of the North Atlantic Ocean. As freshwater is lighter than saltwater and wants to remain on top, the surge halted the sinking of surface waters that forms part of the so-called conveyor belt of ocean circulation, a large-scale, ribbon-like current system that moves warm subtropical surface waters northward, warming the North Atlantic and neighboring regions (the well-known Gulf Stream is part of it). With the shutdown of the warm current, the extratropical North Atlantic and neighboring regions of eastern North America and western Europe cooled down, almost back to ice age–like conditions.[11]

If you've seen the film *The Day After Tomorrow*, then you're familiar with the scenario—or at least a caricature of it. The movie is premised on a dramatically sped-up, exaggerated version of the actual phenomenon. Global warming won't cause monster tornado outbreaks to destroy Hollywood, supercooled plumes of air to freeze people in their tracks, or an ice sheet to re-form over the United States in a week. But there's a grain of truth to the tale. Climate models indicate that human-caused warming could lead to a slowdown in the ocean conveyor. Though there is no ice sheet over North America today, there is one over Greenland, and it is losing ice ever more rapidly. It could one day release enough freshwater into the North Atlantic Ocean to cause a similar, if less dramatic, conveyor shutdown. Some of my own research, in fact, suggests that such a shift is already underway. My collaborators and I showed that a small region of the

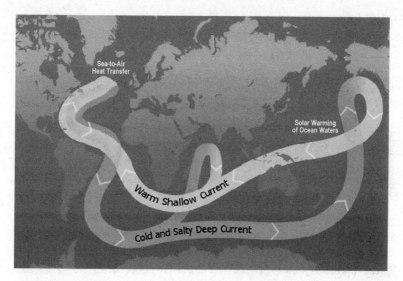

Figure 4. The major surface and deep-water circulation components of the ocean that combine to form the global ocean conveyor belt.

North Atlantic just south of Greenland has actually cooled over the past century, even as the rest of world has warmed. That pattern of regional cooling bears the "fingerprint" of the initial stages of a slowing ocean conveyor.[12]

Scientists have called this return to glacial conditions in the North Atlantic 12,900 years ago the Younger Dryas: "younger" because it's the more recent (and more pronounced) of two cold events that occurred toward the end of the last ice age, and "dryas" after the tundra wildflower (*Dryas octopetala*) that is prevalent in high-latitude lake sediments dating to this time. The conveyor shutdown and consequent cooling of the North Atlantic and neighboring localities lasted roughly a thousand years, before the final exit from the last ice age.

That event seems to have spurred one of the key innovations that would ultimately make human civilization possible. The end of the Younger Dryas 11,700 years ago marked the beginning of the Holocene interglacial (warm) epoch. It also marked the beginning of what archeologists call the Neolithic period, the final stage of the Stone

Age. It has been said that the Stone Age didn't end for want of stones, and that is undoubtedly true. It ended because something better came along. That something was what has been called the Neolithic Revolution, a set of remarkable, interrelated human innovations that allowed for a shift away from our prior nomadic existence toward permanent settlements, agriculture, and farming.

It all appears to have begun in the boomerang-shaped region bordering the eastern Mediterranean Sea known as the Fertile Crescent. The Fertile Crescent spans modern-day Iraq, Syria, Lebanon, Palestine, Israel, Jordan, and Northern Egypt, together with the northern region of Kuwait, southeastern region of Turkey, and western portion of Iran. It is considered to be the "cradle of civilization" because of the numerous technological innovations—including writing, the wheel, agriculture, and irrigation—that all arose there.

Of special importance was the Natufian culture of southern Syria, which developed at the end of the archeological Paleolithic era around 15,000 years ago as we were beginning to exit the Ice Age. What's unusual about the Natufians is that they were largely sedentary rather than nomadic, with small settlements that took advantage of the nutrient-rich soils and relative warmth and rainfall in the Fertile Crescent at that time. These favorable climate conditions allowed for stationary subsistence hunting and gathering. The Natufians had access to abundant wild cereals, legumes, almonds, acorns, and pistachios while hunting gazelle. They used sickles with flint blades and stone mortars and pestles for harvesting and grinding grains. These were precursor tools of agriculture, but the Natufians were hunter-gatherers not farmers. At least not yet.[13]

The Younger Dryas caused a cooling and drying in the Fertile Crescent, which created hardships for the Natufians. With conditions no longer as favorable for sedentary hunting-gathering, they were forced to adopt a variety of adaptive strategies. One group of Natufians, in the Negev Desert and northern Sinai Peninsula, became nomadic, eventually giving rise to the Harifian culture, known among other things for their Harif point arrowheads. Other Natufian groups, however, adopted the opposite strategy, remaining sedentary while intensifying their hunter-gatherer activities. It is this pathway that would lead to cultivation and agriculture.[14]

The Natufians were already aware, from their sedentary foraging, that planted seeds could later take root. That knowledge laid the groundwork for cultivation of wild plants, which appears to have begun during the Younger Dryas in the northern Natufian communities. These communities learned to cultivate various cereal crops as a way to intensify their procurement of food. Those Wheat Chex you ate for breakfast? Thank the Natufians.[15]

The newly acquired agricultural knowledge quickly spread throughout the Natufian communities of the Fertile Crescent. The widespread appearance of green beads as body decorations among Late Natufian communities seems to underscore the importance that cultivation had taken on for their culture. As the Younger Dryas came to an end, the region once again became warmer, with more reliable winter rains. Farming became more widespread and more productive. Animal husbandry also took hold, as hunting and herding naturally transitioned into the domestication of cattle, goats, sheep, and pigs by sedentary farmers.

Meanwhile, something similar was playing out in China. Once again, the climate influences of the Younger Dryas appear to have played a critical role. The warmer and wetter conditions in northern China during the Bølling-Allerød interstadial were favorable for hunter-gatherer communities in the region, and they grew in size and spread into previously arid and subarid environments. But the drier and cooler conditions that set in during the Younger Dryas made foraging more challenging, so they transitioned to the cultivation of wild millets as a means of food procurement. Livestock, such as pigs, were bred as hunters became farmers. The emergence of rice cultivation in southern China may have arisen among foragers in the Yangtze River basin around this time. The adoption of agriculture occurred independently and nearly simultaneously in disparate regions, underscoring how large-scale climate changes drove human societal innovation.[16]

Civilization Is Born

Our species, *Homo sapiens*, had finally made the transition from nomadic to sedentary existence. We had learned to cultivate food crops and raise livestock. We were beginning to transform Earth's natural landscapes. As we became more efficient at procuring food, our numbers increased exponentially. Communities turned into villages and towns, and then cities; social hierarchies began to form. The seeds of civilization had been firmly planted. But it would take several millennia for them to germinate into the form of *city-states*—cities with surrounding territories that formed an independent state, a rudimentary civilization.

Eleven thousand years ago, just 700 years after the end of the Younger Dryas, a Natufian proto-city, Jericho, emerged in the semiarid rolling hills of the Jordan River valley, at the mouth of a spring, east of modern-day Jerusalem. By 8500 BP (before the present), it had been abandoned and then reoccupied, and taken the form of a mud-brick agricultural village, containing a massive stone tower, all of it surrounded by its famous stone wall. There is evidence of domestication of goats as a dependable source of meat during this time. Jericho may have supported a population of several thousand, and the feats of construction, like the tower and wall, suggest the beginnings of coordinated labor. A number of other similar agricultural communities had been established by this time.

Then suddenly, around 8200 BP, we see the large-scale abandonment of many of these settlements in favor of smaller, more self-sufficient, individual household units, as the region once again experienced an abrupt drying event—one that lasted for about two centuries. The paleoclimate evidence suggests that this was a sort of mini–Younger Dryas event, a final massive meltwater pulse as the last vestiges of the Laurentide Ice Sheet melted away. Possibly another ice dam collapsed, releasing into the North Atlantic Ocean sufficient enough pooled meltwater from the residual Lake Agassiz to once again weaken, if not collapse, the ocean conveyor, cool the North Atlantic, and dry out the Fertile Crescent.[17]

Around 8000 years ago, the return to wetter conditions promoted settlement of the fertile valleys of the Tigris-Euphrates river system in the northern Fertile Crescent, where river levees and seasonal water basins were likely exploited for a very primitive form of irrigation by a farming culture known as the Halaf who inhabited the region. Including parts of southeastern Turkey, northern Iraq, and Syria, the iconic role played by this region has earned it a name: Mesopotamia, which in Greek means "the land between the rivers." Villages such as Jarmo were exemplars of this culture, showing evidence of significant technological innovations such as decorated pottery, which replaced more primitive basketry and goat skin bags. Much of the diet came from the farming of wheat and barley and the raising of domesticated goats. The size, shape, and configuration of individual homes suggests that the primary social unit was now the extended family, while large grain storage pits suggest cooperation at the community level. Primitive social hierarchy was now necessary to determine who gets what land and how it is passed along generationally. Somebody—an individual or group—was needed to make those decisions. That encouraged more elaborate sociopolitical systems, greater societal complexity, and power hierarchies. Jarmo, for example, has been interpreted as an early chiefdom.[18]

Roughly 2000 years later, around 6000 years ago—what is called the mid-Holocene—the first true human civilizations—city-states—emerged in Mesopotamia. We can once again thank climate change. To understand how and why, we'll need to consider another cyclicity involving Earth's orbit around the Sun. This one has to do not with the shape of Earth's orbit, but rather the *precession of the equinoxes* (the wobbling of the orbit). This cycle occurs over shorter periods of time—not 100,000 years but closer to 26,000 years—and the impact on the tropics and subtropics is especially pronounced.[19]

The axis of Earth's rotation isn't oriented at a right angle relative to the plane of Earth's orbit around the Sun. It tilts at about 23.5 degrees from vertical. And Earth wobbles like a top that has been spun. Only it takes about 26,000 years for a single full wobble back and forth. Today, the wobble orients the Northern Hemisphere so that it is tilted away from the Sun (at the winter equinox, which is December 22) around the same time we're closest to the Sun (at the perihelion, which is currently January 4). Because the Northern Hemisphere is closer to the Sun during winter, its winters are a bit less cold. Similarly, the Northern Hemisphere is farther from the Sun in mid-summer, making summers a bit cooler. Seasons are generally milder. Twelve thousand years ago, at the beginning of the Holocene, it was nearly the opposite situation, and seasonality was enhanced.

And, most relevant to our story here, 6000 years ago—during the mid-Holocene—we were halfway in between the two extremes, slowly transitioning toward reduced seasonality.

Land heats up faster than the ocean. So, in coastal regions, we get sea breezes in the summer in the late afternoon when the land heats up and the air rises, drawing in cooler, moister air from the ocean. A monsoon can be thought of as a continental-scale sea breeze circulation, and when seasonality is greater and summers are warmer, you get stronger monsoons. The Indian summer monsoon, which we discussed earlier in the context of longer-term tectonic impacts on the climate, is one example, but there is a related summer monsoon in West Africa.

Both appear to have played a role in what transpired in Mesopotamia. Climate model simulations and paleoclimate proxy observations such as geochemical evidence from cave deposits indicate that the region was wetter in the early Holocene owing to a combination of a stronger and more northerly South Asian summer monsoon, bringing more Indian Ocean moisture to the region, and a stronger West African monsoon, which led to greater transport of Atlantic Ocean moisture into North Africa and a greening of the Sahara Desert. Increased vegetation in the Sahara may have altered wind patterns in a way that favored greater movement of moisture into Mesopotamia. By the mid-Holocene, as seasonality was weakening, so were the monsoons. The region thus transitioned toward a semi-arid climate.[20]

Much as the adverse shift in climate during the Younger Dryas put selective pressure on coping strategies that resulted in the development of cultivation and agriculture, the shifting climate in Mesopotamia by the mid-Holocene put pressure on the efficient use of water. The "land between the rivers" was ideally suited to irrigation. What was needed was the societal organization to implement a far more sophisticated version of it, one that embraced basic hydraulic engineering principles. City-states could provide that organizational support.[21]

Thus was born civilization in Mesopotamia. The first city-state was Sumer, established in the south (what is now south-central Iraq) and permanently occupied by the beginning of what is called the Uruk period (6100–4900 BP). Aided by irrigation technology, farmers here were able to grow an abundance of grain and other crops. That freed other citizens up to perform other tasks such as construction, leading to the formation of the first true urban settlements.

Sumer was divided into numerous independent city-states, each with a population exceeding 10,000 people. The independent states were separated by canals and stone boundaries. There was trade between them, with goods typically transported along the canals and rivers of southern Mesopotamia, which in turn led to greater homogeneity of economies and cultures. The absence of walls suggests a general absence of warfare among the different states, though—as we are now into the Copper Age—there are knives, drills, wedges, saws, spears, bows, arrows, and daggers. The latter suggest at least occasional battles.

There were numerous common features that emphasize the increasingly "civilized" nature of these city-states, though we also see here some of the foibles that would be inherited by later societies. Each state was theocratic, centered on a temple dedicated to a patron god or goddess of the city. Centralized administrations employed specialized workers and, alas, later slaves for labor. The social structure was male-dominated and stratified, headed by a priest-king, who was assisted by a council of elders. There was law—though it was sometimes brutal and unjust. Women found guilty of unfaithfulness were stoned to death with rocks. Among the initial city-states were Uruk, Ur, and Akkad. Uruk became the most urban city the world had yet seen, with a population exceeding 50,000, comparable to a small modern city.

Many of the features we today associate with civilization were now evident. There was elaborate pottery in the form of vases, bowls, and dishes. There were jars for honey, butter, oil, and even early wine (probably made from dates). They were sealed with clay. Individuals wore headdresses and had necklaces made from gold. There was rudimentary furniture, including beds, stools, and chairs. They had fireplaces and altars. They had written language, using tablets for writing. And they played music with lyres and flutes.

Other civilizations arose elsewhere around the same time in other subregions of the semi-arid Mediterranean and Middle East, drawing in each case upon sophisticated irrigation projects to deal with challenges of a drying climate. We see it in Ancient Egypt around 5000 BP, the Indus Valley Civilization around 4500 BP, and the Minoan civilization around 4000 BP. Climate change, once again, seems to have fostered similar innovations independently and nearly simultaneously in disparate regions.

But something unique happened in Mesopotamia. By 4300 BP, Sumerian and Akkadian speakers would unite under one rule and one ruler, Sargon of Akkad. The world's first true empire was born. Centered in the city of Akkad, the Akkadian Empire extended to a vast surrounding region, using its formidable military might to exercise influence throughout Mesopotamia and neighboring regions, including Anatolia and Saudi Arabia.

The social organization, division of labor, and social hierarchy of an empire provided increased power and influence—in the form of a soldier class and increasingly advanced and effective weaponry crafted by metalworkers. It also provided increased resilience, in the form of sophisticated irrigation projects that could support farming even when rainfall became increasingly unreliable and intermittent as the region continued to grow more arid. But resilience has its limits. And so we come to perhaps the most instructive lesson of all here: the fall of the Akkadian Empire around 4200 BP.

Now, we must be wary of climate determinism: the notion that every significant historical event, every societal origin or collapse, can be interpreted entirely through the lens of climate change. We must always appreciate the complexities of human behavior and sociopolitical dynamics that effect societal changes. That being said, it is likely that an abrupt climate shift, and its interaction with societal dynamics, was *the* fundamental factor behind the fall of the first great empire.[22]

This hypothesis was first put forward by Yale anthropologist Harvey Weiss in the early 1990s based on his detailed studies of the archeological remains of Tell Leilan, which had been the administrative center of the Akkadian Empire. (Highly controversial at the time, there is now considerable paleoclimate evidence confirming the hypothesis.)[23]

The precise cause of the drought is debated. It might have been a large volcanic eruption. An explosive tropical volcanic eruption, like the Year Without a Summer eruption of Mount Tambora in Indonesia in 1815, could eject enough particulate matter into the stratosphere to block out a significant amount of sunlight. Though nothing compared to the asteroid impact that chilled the planet and killed the dinosaurs sixty-six million years ago, it could be enough to cool off and dry out the subtropics for more than a decade. Sediment deposits analyzed recently by one group of geologists indicate an eruption of the Cerro Blanco Volcanic Complex in the tropical Andes of northern Argentina around 4200 BP that was probably one of the most explosive eruptions of the Holocene epoch.

Other civilizations in the region were also impacted by this pronounced, widespread drought. The Old Kingdom of Egypt, builders of the majestic pyramids, as well as the Indus Valley (or Harappan) Civilization and the Early Bronze Age civilizations of Palestine, Greece, and Crete all saw diminished agricultural production. Yet only the Akkadian Empire underwent immediate collapse, seemingly because of the challenges of keeping such a sprawling and diverse civilization united.[24]

The Akkadian Empire had become dependent on the productivity of the northern part of the empire, using their bountiful agricultural yields to distribute food to other regions and to support their massive army. The devastating consequence of the drought was encapsulated by "The Curse of Akkad" text, which gave a decidedly stark assessment of the predicament: ". . . the large arable tracts yielded no grain, the inundated fields yielded no fish, the irrigated orchards yielded no syrup or wine, the thick clouds did not rain." The agricultural collapse was followed by mass southward migration of the northern populations, which was met with opposition from the local southern populations, including the construction of a hundred-mile-long wall extending all the way from the Tigris to the Euphrates to keep out the immigrants. If that scenario sounds disturbingly familiar, there's a reason, which we'll get to later.

Though other civilizations in neighboring regions didn't undergo imminent collapse, they were adversely impacted by the drought, and within a matter of centuries they, too, would meet their demise in substantial part because they had become dependent on trade with Mesopotamia that evaporated with the collapse of the Akkadian Empire. The Indus Valley Civilization would disappear within a few hundred years. Then there's the Minoan civilization. I've had the pleasure of witnessing—and indeed consuming—their magnificent cultural achievements, having visited excavation sites in Crete and Thera (aka Santorini) and having imbibed a rather tasty varietal of white wine known as Assyrtiko that originated with the Minoans. Minoan civilization disappeared by 3500 BP, hastened by a destructive volcanic eruption in Thera. If you ever get the chance to swim in the hydrothermally warmed waters of the Nea Kameni caldera that was left behind, I recommend it.

Though we've focused on Europe and Asia thus far, a remarkably similar scenario played out in the Americas. The continent was peopled between 15,000 and 17,000 BP when Asian tribes crossed the land bridge to North America, exploiting the low sea level stands during the late stage of the Ice Age. Within a few thousand years, some traveled all the way down to Peru. And it was there, once again in the mid-Holocene, when the first true American civilization—the Caral—arose. Climate change was once again a key factor, in the form of a phenomenon known as El Niño.

El Niño is the periodic warming of the surface waters of the eastern tropical Pacific Ocean that today happens every three to five years. (It's called El Niño—"the child"—after the Christ Child, because the warming always tends to emerge around Christmas time.) It's tempting to say that the warming results from the weakening of the trade winds (the surface winds that tend to blow from east to west in the tropics), but causality is a tricky matter here. Strong trade winds cause deep cold waters to rise to the surface (known as upwelling) in the eastern equatorial Pacific, cooling the ocean surface. Warmer waters are found farther west on the equator over Indonesia. Those warm waters heat the atmosphere, making it rise. The warm air then travels east aloft along the equator, sinking back to the surface in the eastern Pacific, and traveling back west along the equator, giving us those very same trade winds that we started with and completing the atmospheric circulation pattern. We can't say that the ocean causes the atmosphere to do what it does any more than we can say that the atmosphere causes the ocean to do what it does. Instead, it's a coupled, interdependent, internally consistent state of the ocean-atmosphere system.

If that state is upset by a weather disturbance that weakens the trade winds, then surface waters warm in the equatorial east, and you lose the contrast with the warm western equatorial surface waters. But it's that temperature contrast that drives the atmospheric circulation pattern responsible for the trade winds, so that entire pattern weakens, and the trade winds weaken further. It's a self-enforcing loop. The system tends to oscillate back and forth every few years between a weak circulation, which favors El Niño, and a strong circulation, which is associated with La Niña, when surface waters are cold in the eastern equatorial Pacific. This entire ocean-atmosphere phenomenon is known as the El Niño/Southern Oscillation, or ENSO. The cooling and warming in the equatorial Pacific Ocean nudge the jet streams in both hemispheres, changing seasonal weather patterns in North America, Africa, and Australasia. El Niño years tend to be wet in western North America with relatively quiet Atlantic hurricane seasons. La Niña years tend to be the opposite: dry out west and stormy in the Atlantic.

Today, El Niño events are common, and we've seen some big ones recently, like the 1982–1983, 1997–1998, and (unusually long) 2014–2016 events. But we've also seen long stretches of cool La Niña conditions in the tropical Pacific Ocean. Will climate change cause bigger and more frequent El Niños? Or might it just lead to the opposite, pushing us toward more of a La Niña–like climate state? The paleoclimate record provides some hints, which we'll get to later on.

The important thing, though, is that we're now in a position to understand the possible reason for a dramatic weakening of the ENSO phenomenon that took place during the early to mid-Holocene according to a host of paleoclimate data, including preserved corals and ocean and lake sediments. The increasing seasonality at this time, as discussed earlier, caused a strengthening of the monsoons. That, in turn, favored stronger trade winds in the tropical Pacific Ocean during the very time—Northern Hemisphere winter—when El Niño events tend to emerge. If the trade winds are too strong, it's difficult to trigger a weakened circulation. In other words, it's difficult for an El Niño to take hold. As the monsoons and trade winds gradually weakened, the long period of dormant El Niño activity came to an end. By 5000 BP, El Niño was ramping up again.[25]

Some of the most biologically rich regions of the world's oceans can be found along the Peruvian coastline, thanks to the coastal upwelling caused by trade winds that pump deep nutrient-rich waters to the surface. The nutrients are used by surface-dwelling phytoplankton, photosynthesizing sea life that support a rich aquatic food chain, including massive populations of sardines, anchovies, and mackerel. El Niño events cut off the upwelling, shutting down this remarkable natural fishery.

The resurgent El Niño episodes consequently led to intermittent interruptions of the key food source for the native coastal Peruvian fishing culture. But while El Niño events were bad for fishing, they were good for agriculture; the warmer coastal waters during El Niño events fed torrential rains in a region that is normally a coastal desert. With water comes the possibility of cultivation of crops and agriculture, but only if you can store the water to get through the long dry periods between El Niño events. That required water storage and irrigation technology. Only the organizational structure of civilization could support such innovations. And thus was born the complex civilization known as the Caral. They engaged in centralized food production through a dual economy of fishing on the coast and agriculture inland, with trade between the two. They built permanent homes, sunken plazas for group gatherings, and eighty-five-foot-high pyramids. The civilization lasted roughly a thousand years. If that seems brief, consider that it is roughly the amount of time that has elapsed between Erik the Red's colonization of Greenland and today.[26]

Despite the disappearances of all of these early civilizations, human civilization itself would survive, thrive, and spread. In Peru the Chavín culture would cultivate maize in the Andean highlands around 3800 BP. Around the same time, the Maya of Central America would cultivate stable crops of maize, beans, squashes, and chili peppers. After the fall of the Akkadian Empire, the people of Mesopotamia eventually coalesced into two major Akkadian-speaking nations: Assyria in the north and Babylonia in the south. The Old

Kingdom of Egypt disintegrated at roughly the same time as the Akkadian Empire, but this was less a "collapse" than a century-long "interlude" as the Old Kingdom transitioned to the New Kingdom. In Greece, the Mycenaean civilization flourished in the Late Bronze Age (3700–3100 BP), followed by the Ancient Greek civilization, which was marked by a number of novel political, philosophical, artistic, and scientific achievements—including the introduction of democracy as a governing system. Those contributions would fundamentally shape Western civilization as we entered the Common Era.[27]

The Common Era

The Common Era (CE), the roughly 2000-year-long era in which we currently reside, begins with year 1 of the Gregorian calendar (I began writing this book in 2022 CE). Within this era we find a number of examples of the subtlety in how climate has impacted human civilization. As we'll see, it's not simply a volcanic eruption or other climate event that wipes out civilizations, but often a complex interplay between climate stresses and sociopolitical factors that leads to societal collapse. There is perhaps no better example than Ancient Greece and Rome.

Greek civilization arguably ended with the death of Alexander the Great in 323 BCE, and was certainly kaput by the time the Romans conquered Greece in 146 BCE. It is noteworthy that, while Greece fell, the Romans actually adopted much of its culture, including its democratic political and social structures and the same pantheon of deities, demonstrating the ambiguity in the notion of downfall. Greek culture, after all, survived. The Roman Empire officially began in 27 BCE when Caesar Augustus declared himself Emperor of Rome. So, give or take twenty-seven years, the Common Era basically begins with the rise of the Roman Empire. The reign of Augustus was characterized by relative peace (known as the Pax Romana) in the Roman world, which was largely free from conflict for more than two centuries, even as it continued with its imperial expansion on the empire's frontiers.

Within two centuries, the Romans controlled an expansive, geographically diverse region of the globe, from northern Britain to the edges of the Sahara and from the Atlantic Ocean to Mesopotamia. Its population enjoyed prosperity and reached a remarkable seventy-five million in number. Now, among climate historians, there is a widespread view that climate-induced challenges were a significant driver of the fall of this great empire. Empire is born, expands, initially thrives, becomes too thinly spread, encounters challenges—probably climate-induced—and falls.

It's not that simple. Though many accounts attribute the fall of the Roman Empire to climate change, this is often based on selective, anecdotal, or tenuous evidence. The argument typically goes that there was a putative Roman Climate Optimum from around 200 BCE to 150 CE, when temperatures were purportedly warmer than today. The Romans were lucky, goes the argument, in having established their empire during this favorable time period. Then, climate conditions, driven by a number of large volcanic eruptions, rapidly deteriorated during the so-called Late Antique Little Ice Age, or LALIA, and the empire fell apart.[28]

It's a good story, fit for an HBO docudrama. But it's not correct. State-of-the-art assessments of the paleoclimate evidence show no evidence of a centuries-long warm period in that region, let alone globally. The evidence instead argues for regionally variable temperature and rainfall patterns and warmth during that time that did not approach modern levels. The argument that volcanic-driven LALIA cooling drove the collapse ignores the fact that the cooling occurred in the sixth century CE, whereas the initial collapse of the empire—the collapse of the *western* Roman Empire (essentially the western Mediterranean and neighboring regions)—occurred a century earlier, in the late fifth century CE.[29]

Societal pressures unrelated to climate change appear to have caused the downfall of the western Roman Empire. The poorer, decaying west became overextended, facing growing unrest from a peasant underclass resentful of the decadent lifestyle of elites. The empire was weakened by repeated invasions, severely damaged in 410 CE when the city of Rome was sacked by the nomadic Visigoths, and finished off in 476 CE when Odoacer, an Italian military leader of Germanic descent, deposed Romulus Augustulus, the last Roman emperor of the west.

But what about the *eastern* Roman Empire (also called the Byzantine Empire)? A collapse did occur there, but nearly a thousand years later, in the fifteenth century CE. The argument for sixth century cooling and drying as being the primary cause is once again highly implausible. But climate change likely *did* play a role. It might have actually *extended* the life of the eastern part of the empire. Paleoclimate proxy data such as lake sediments and stalactite and stalagmite deposits (called speleothems) indicate a trend toward wetter winters in the eastern Mediterranean beginning in the fifth century CE. Winter was the wet season in the region, and winter rains (and snowpack in mountainous regions) were, as they are today, critical for recharging water supplies that must hold through the long Mediterranean summers—summers that have become increasingly dry in recent millennia as part of the mid- to late Holocene drying trend discussed earlier.

What might have been behind the shift toward wetter winters that extended the life of the Byzantine Empire? Winter rainfall in the eastern Mediterranean and Middle East is dominated by the position of the jet stream as it crosses the Atlantic Ocean from west to east. Some years it heads north, bringing warm and stormy, wet conditions to Europe, but denying rainfall to more southern latitudes. Other years, it heads straight across the Atlantic into the Mediterranean region, bringing ample storm rainfall with it. This fluctuation in jet stream behavior from year to year is called the North Atlantic Oscillation, or NAO, and the latter configuration is its negative phase, associated with cold and dry conditions in Europe and wet conditions in the Middle East. The NAO is influenced by changes in solar heating. Though the mechanisms are somewhat complicated, the net effect is that a downturn in solar heating favors this negative NAO phase. Such a downturn documented in radiocarbon deposits in ice cores began in the mid-fifth century CE. It would likely have led to a negative NAO pattern and the increased rainfall that occurred in the eastern Mediterranean and Middle East at that time.[30]

This climate trend clearly played a favorable role in the eastern part of the Roman Empire. Evidence from pollen deposits and historical sources reveal that farming in the region thrived in the mid-fifth century CE, with an expansion of agricultural settlements. The taxation system used by the empire would have benefited from the increased agricultural productivity, providing more support for critical infrastructure, including sophisticated irrigation systems and resilience-favoring hydrological projects, further insulating farmers in the most arid subregions, all of which would support further expansion and consolidation of the empire.[31]

Byzantium did eventually fall in the Middle Ages. It had been weakened by the plague, with nearly half of the population of Constantinople perishing from the outbreak of 1347–1353 CE (and yes, climate may have played a role in the spread of the disease from Asia to Europe). The empire would continue to weaken and fragment, ending with the fall of Constantinople to the Ottoman Empire in 1453.[32]

Climate change would continue to play at least some role in guiding human history in subsequent centuries. Preindustrial temperature changes were modest at the global scale. The so-called Little Ice Age, broadly defined as the fifteenth to nineteenth centuries, appears to have been less than 2°F cooler than the previous four centuries globally—that small difference attributable to modest changes in natural drivers of the climate (solar heating and volcanic eruptions). Regional shifts in temperature and rainfall were largely, however, the result of changes in atmospheric circulation related to the NAO and ENSO, which may themselves have resulted from those same natural drivers. The medieval era (eleventh to fourteenth centuries) was relatively warm over large parts of North America, the North Atlantic Ocean, and Eurasia, but cool in the tropical Pacific Ocean. It was especially dry in the western United States ostensibly due to a persistent cool La Niña pattern in the tropical Pacific. These sizeable regional changes in climate certainly impacted human history and likely contributed to at least some societal collapses.[33]

The canonical example is the demise of the Norse settlements of Greenland during the fifteenth century. In 985 CE, Erik Thorvaldsson (Erik the Red) led a fleet of ships traveling from Iceland to southern Greenland, carrying around 500 men and women, livestock, and supplies. They maintained permanent settlements, raised dairy cattle and sheep, and grew grain, while maintaining, through able seafaring, trade with mainland Europe. Then it all fell apart in the fifteenth century.

The Norse had benefited from relatively warm conditions thanks in part to the positive NAO pattern that persisted during medieval times. As the positive NAO pattern gave way to a more negative NAO in the fifteenth century, the chill set in. The cooling certainly presented a challenge to these hunting and farming colonies, plausibly at least contributing to the demise of the Norse settlements between 1400 and 1450 CE. But poor decision-making contributed to the collapse, too. Scientists have argued, for example, that the overreliance of the Norse on walrus tusks, which was a valuable but dwindling trade commodity, played a key role. Additionally, the collapse of trade with mainland Europe, which was impeded by increasingly thick sea ice in the region, was at least an indirect consequence of the cooling climate, showing how climatological and sociopolitical factors combined to collapse a civilization. The demise of the Norse colonies undoubtedly altered the course of history. It opened up the opportunity for the Dutch to emerge, as we saw in the introduction, as the dominant seafaring nation. Loser, meet winner.[34]

The poster child for climate collapse in North America is found a number of centuries earlier with the maize-farming civilization known as the Anasazi that inhabited the Four Corners region of the American Southwest beginning in the twelfth century BCE. The civilization ended dramatically around 1300 CE. Anasazi agriculture, it appears, could not withstand the increasingly dry conditions in the western United States at that time. The drought came to a peak locally in the late thirteenth century CE, resulting in the abandonment of Anasazi settlements. In yet another example of climate-induced stress interacting with societal decision-making, scientists have used a type of modeling known as agent-based modeling to simulate how the Anasazi might reasonably have responded to changing climate conditions by altering their crops and settlement locations. Though actual Anasazi civilization collapsed rapidly and dramatically at 1300 CE, the modeling suggests that agriculture should have remained viable in the simulated civilization, albeit at a reduced level, well after that, underscoring the sometimes-complex relationship between climate and collapse.[35]

In some cases, the impacts of climate change, and even the question of whether they were adverse or beneficial, are ambiguous. There are numerous accounts, for example, of how the colder temperatures of the Little Ice Age in Europe—where cooling was more pronounced (thanks to the prevalence of a negative NAO weather pattern that ushered in cold air from the Arctic)—had negative societal impacts. Especially cold conditions in the late seventeenth to early eighteenth century CE in France, for example, appear to have led at least initially to poor harvests, spikes in food prices, and widespread mortality. But the French government engaged in adaptive strategies to mitigate hunger and death, by importing grain from Algeria, for example. And Londoners celebrated the winter cold with frost fairs on the River Thames. The Dutch were able to avoid food shortages by diversification of diets and actually benefited, as we learned earlier, by taking advantage of stronger winds to shorten their ocean voyages, improving trade with Asia.[36]

Another case in point is the French Revolution. Some accounts ascribe climate-driven factors to the event, citing the Little Ice Age as an underlying causative factor but also implicating other climate impacts, like consecutive years of heat and drought in 1788 and 1789 and even a severe hailstorm on July 13, 1788, all of which served to decimate crops and impair the food supply. As one writer described it, "the years of climatic stress, financial instability and political conflict brutally converged in 1788 and 1789" to yield the revolt. The problem is that there's actually no evidence that the Little Ice Age—a period of *colder* conditions in Europe—would have caused episodes of extreme *heat* (and drought) in France, let alone freak individual weather events like hailstorms. Sometimes, weather is just weather, not climate.[37]

The underlying causes of the French Revolution were largely political rather than environmental. The prevailing government failed to deal with issues of social and economic equality (sound familiar?) and did a poor job managing the economy, the result of which was widespread unemployment, economic depression, and spiking food prices. Add in working class resentment over the extravagant lifestyles of the ruling elite, and you've got all of the ingredients necessary for a revolution. Could climate-related stresses have been an aggravating factor? Yes. Can they be *blamed* for the French Revolution? No.

Examples of the complexity of societal collapse are not limited to Europe and North America. Consider the fall of the Mayan civilization of southeast Mexico and Guatemala. A series of major, extended droughts in the ninth and tenth centuries CE (recorded in chemical deposits in ocean sediments) purportedly led to the collapse of the classic Mayan civilization in the tenth century CE. Yet the one-by-one disappearance of lowland Mayan cities was already underway in the seventh century. Documented evidence of drought is not consistent in either space or time with the demise of Mayan cities, many of which experienced only minor disruption and even flourished until the Spanish conquest of the early sixteenth century CE.[38]

Cautionary Tales

From sixty-six million years ago, when our distant rodent-like ancestors crawled out from the shadows of the dinosaurs, to five million years ago, when our less-distant primate ancestors came down from the trees to hunt on ancient African savannas, climate has shaped us. The coming and going of ice ages during the past several hundred thousand years turned us from *hominins*—apes—into *hominids*—humans—as the challenges of surviving huge swings in climate favored bigger brains and greater intelligence. A cold blip at the end of the last ice age 12,000 years ago forced us to learn how to cultivate plants and farm.

Since then, individual civilizations have repeatedly risen and fallen, and climate change has frequently been implicated in both instances. Yet human civilization as a whole has been remarkably stable: we have thrived and our numbers have increased exponentially. It's not a coincidence that our stability as a species mirrors the stability in global temperatures over this precise timeframe. This interval of stable global climate has been punctuated by volcanic eruptions, solar fluctuations, and slow shifts in regional rainfall and temperature patterns, monsoons, and the El Niño phenomenon due to the precession of the equinoxes. But sizeable shifts in climate have been limited to regional scales, allowing people to migrate to new, more favorable areas when conditions deteriorate locally. The changes in climate today are driven by fossil fuel burning, and they are global. No place is safe from the detrimental impacts—coastal inundation due to global sea level rise and stronger storms, drying continents, and more extreme weather events.

The fact that global temperatures were so stable over the past 6000 years itself demands explanation. A compelling explication of this mystery has been put forward by paleoclimate scientist William Ruddiman, whose work we encountered earlier. Ruddiman sought to understand the mystery of why the global climate has been so stable over the past 6000 years when it should have been slowly cooling, descending ever so gently into the next ice age. He summarized his arguments in his 2005 book, *Plows, Plagues, and Petroleum*—we'll get to the relevance of each of the words in that alliterative title shortly.[39]

Ruddiman is now retired from the University of Virginia, working from home in a bucolic sanctuary in the Shenandoah Valley of Virginia, perched on a foothill, nestled between the Appalachians to the west and the Blue Ridge Mountains to the east. I had an opportunity to discuss this work with him shortly before I wrote this chapter, while visiting him at his place mid-March 2020 on my way to the Virginia Festival of the Book in Charlottesville. Controversial at first but backed up by years of painstaking work, he has made a compelling case that we humans didn't take control of the climate during the industrial revolution of the past two centuries. We actually took control more than 6000 years ago—through agriculture, deforestation, and the spreading of plagues.

Compared with past glacial/interglacial cycles, Earth, Ruddiman argues, should have begun the slow but inexorable transition from warm interglacial to cold glacial conditions about 9000 years ago, with dropping CO_2 levels, cooling temperatures, and more extensive ice cover. Given the difference of roughly 9°F between peak interglacial and peak glacial conditions over the course of the roughly 100,000-year eccentricity cycle, the climate should have cooled about 1°F over this 6000-year timeframe. But that's not what happened. Global temperatures were almost perfectly flat. The solution to the riddle, Ruddiman argues, was that this natural cooling trend was offset by modest, but non-negligible, human-caused greenhouse warming.

Clearly there weren't coal-fired plants or SUVs thousands of years ago. But increasingly widespread deforestation for slash-and-burn agriculture in Eurasia and North America released increasingly large amounts of CO_2 into the atmosphere (the "plows"). At times, CO_2 actually dipped a bit due to drops in population resulting from pandemics, such as the bubonic plague in Europe during the Middle Ages and the spread of smallpox through Native American populations following European colonization of the Americas in the late fifteenth century (the "plagues"). But the overall pattern over the past 6000 years has been a steady, long-term rise in CO_2 levels when orbital changes should have been driving them down. I find this to be a compelling argument, based on the evidence we have.

Then there's methane. As we learned earlier, rice cultivation in China began in the lower Yangtze River valley of China around 6000 years ago, and it quickly spread to other parts of China and then elsewhere in Asia over the ensuing millennia. Rice farming requires flooded paddies, which creates extensive pools of standing anoxic water—a perfect breeding ground for methanogenic (methane-generating) anaerobic bacteria. Methane, though it has a shorter atmospheric lifetime than CO_2 because it oxidizes rapidly in the atmosphere, is a very potent greenhouse gas. As long as it continues to be generated, it continues to have a warming impact. Ruddiman links the anomalous long-term rise in methane (which, too, should have been in decline) to the rise in rice cultivation.[40]

Now we appreciate the potential role played by both "plows" and "plagues." And "petroleum"? Oil as well as coal and fossil gas are the fossil fuels that we've been burning since the dawn of the industrial revolution. That has caused a far more rapid and prodigious spike in atmospheric CO_2 concentration. And natural gas extraction through hydraulic fracturing (aka fracking) is releasing increasingly large amounts of "fugitive" methane into the atmosphere. It seems we've become increasingly adept over time at turning up the thermostat.

Thus, we return to the ironic twist with which we began this chapter. Climate has indeed shaped us. Our brainy species arose in part because of the selective pressures of having to deal with the huge swings between glacial and interglacial climates that arose in the late Pleistocene. And we ultimately used our brains to create civilization, to develop irrigation and agriculture, and to build city-states. These same activities warmed the planet, just a bit. Just enough, in fact, to offset the slow descent into the next ice age. We had achieved the ultimate adaptation: the ability not simply to be controlled by climate but to actually control climate. We helped create our *fragile moment*, a stable global climate upon which to build the infrastructure of human civilization.

We should have stopped while we were ahead. But we went further. We constructed an *industrial* civilization that was entirely dependent on fossil fuels. We used our intelligence and ingenuity to generate energy by mining and burning coal, oil, and fossil gas. That energy in turn aided industrial-scale agriculture. These activities helped us support a population of eight billion people on this planet. But they began to warm the planet not a little but a lot, altering the global climate, shifting rain belts, expanding deserts, melting ice, raising sea levels, and unleashing devastating extreme weather events.

In the time span of just a couple centuries, we developed technology that takes millions of years of buried carbon and returns it to the atmosphere in a geological instant. Quoting again the great Carl Sagan: "Our civilization runs by burning the remains of humble creatures who inhabited the Earth hundreds of millions of years before the first humans came on the scene. Like some ghastly cannibal cult, we subsist on the dead bodies of our ancestors and distant relatives."[41]

The industrial civilization we created has clearly engendered challenges, the climate crisis foremost among them. That is in part the peril Sagan spoke of in the quote that opens this chapter. But it has also afforded us some degree of resilience and even opportunity, insulating us to an extent from the effects of climate variability and climate change. That is in part the promise Sagan spoke of. We have sophisticated technology today that we can employ in an effort to adapt to climate change. We can build coastal defenses against sea level rise, adapt crops and cultivars to shifting temperature and rainfall patterns, and manage water resources and agriculture in the face of longer dry seasons and worsening droughts. Most importantly, we have the technological know-how to decarbonize the global economy, moving away from the harmful burning of fossil fuels toward clean energy and climate-friendly agricultural and land use policies. The obstacles here aren't technological. They are political.

We also have distinct advantages over the past civilizations we've studied in this chapter because, unlike them, *we have the ability to anticipate the future.* Although far from a perfect crystal ball and subject to uncertainties, climate models provide a road map for how the climate system is likely to evolve in the future. Moreover, they inform our understanding of how rapidly we must reduce carbon emissions to avert dangerous levels of climate interference. We can also use demographic model projections to estimate where population growth will be greatest. All of that information can be employed to design strategies that minimize the impact of climate change on the societies that are at greatest risk. But we must recognize that there are thresholds beyond which we will simply exceed the adaptive capacity that human civilization affords us. Here again, we must respect the lessons from past examples of societal collapse.

The collapse of both the Anasazi civilization and Akkadian Empire detailed earlier in this chapter serve as cautionary tales. As we

Figure 5. Cliff Palace in Mesa Verde National Park in Colorado was abandoned in the late thirteenth century due to the impacts of extreme drought on the Anasazi civilization in the Four Corners region of the desert Southwest.

saw, modeling that takes into account climate conditions does not predict the observed Anasazi collapse in 1300 CE, suggesting the possibility that our models are missing something. It's possible that we are underestimating the fragility of civilizations when subject to climate-driven stresses. That our best modeling efforts seem to under-predict the potential for societal collapse, in this key test case where we know precisely when it occurred, should give us pause when it comes to the unprecedented and uncontrolled planetary experiment we are now running. If our models tend to underpredict key impacts, then where there is uncertainty in the results of an intervention, we should err on the side of caution (this maxim actually has a name, the precautionary principle). And the stakes couldn't be greater to-day—the continued burning of fossil fuels threatens the very viability of human civilization.

The Akkadian Empire offers us multiple lessons as its echoes from millennia past reverberate today. There is the matter of the wall known as the "Repeller of the Amorites" built from the Tigris to the Euphrates by the Akkadians in a desperate effort to keep out immigrants as climate conditions deteriorated. It is difficult not to draw a connection with another wall: the wall that former U.S. president Donald Trump and his acolytes wish to construct at the southern border of the United States to keep out Mexican and Central American refugees. There are multiple factors behind the ongoing exodus. Among them are a desire to flee poverty and violence. But human-caused climate change, through its impact on food security, is certainly an underlying factor. Speaking about this mass migration, Andrew Harper, an adviser to the United Nations High Commissioner for Refugees, argued that "climate change is reinforcing underlying vulnerabilities and grievances that may have existed for decades, but which are now leading to people having no other choice but to move."[42]

Then there's what's happening today in modern-day Mesopotamia itself. Unrest erupted in Syria in mid-March 2011, part of the larger uprising that has come to be known as the Arab Spring. The unrest grew into an outright civil war in Syria that has now cost hundreds of thousands of lives and has produced one of the greatest mass migrations on record. The underlying cause was a decade-long drought in Syria that is likely the worst in at least a millennium. The unprecedented drought—exacerbated, if not outright caused, by climate change—decimated agriculture in the region and forced rural farmers into the cities of Aleppo and Damascus, competing for food, water, and space with existing residents. The resulting conflict, unrest, and violence created an ideal environment for terrorist organizations to recruit disaffected individuals, and thus was born the international terrorist organization known as the Islamic State of Iraq and Syria, aka ISIS.[43]

The most important lesson of all, however, is that large civilizations are both resilient and fragile at the same time. They are like catamarans. The double hull design of a catamaran makes it very stable with respect to moderate forces from waves or wind tending to induce it to roll. But the very design that provides great stability for small forces leads to total instability—a literal tipping point—for large ones. The Akkadian Empire was able to reduce its vulnerability to limited water resources using the tools of civilization: large work forces that could implement water storage and irrigation and the transportation of resources from where there were surpluses to where there were deficits. But sprawling civilizations are fragile, requiring cooperation and a degree of common interest among diverse constituencies. In the face of a large force that took the form of an epic drought, the empire collapsed. What implications does that have for our truly globally connected, planetary-scale civilization today? Is it susceptible to collapse given a large climate perturbation? More precisely, just how large a perturbation can we endure?

In the chapters that follow, this will be the overriding question. We will look at episodes of climate change in the past and learn about the response of our climate system to disruptions both big and small. We will start at the very beginning, with the primordial Earth and two specific episodes that occurred early in Earth's four-and-a-half-billion-year history: the Faint Young Sun and Snowball Earth. These two episodes speak to a seeming contradiction—the simultaneous resilience and fragility of the very climate system upon which modern human civilization is reliant.

2

Gaia and Medea

Snowball Earth and the Faint Young Sun

Some say the world will end in fire,
Some say in ice.
From what I've tasted of desire
I hold with those who favor fire.
But if it had to perish twice,
I think I know enough of hate
To know that for destruction ice
Is also great
And would suffice.

—ROBERT FROST, *"Fire and Ice"*

This remarkably stable 6000-year period during which human civilization has flourished has benefited from stabilizing feedbacks that help maintain a climate "happy medium." You can push the climate system a little, and it doesn't go off a cliff. Instead, over time, it settles into a slightly different climate state. But if you push hard enough, it can spiral out of control. As we continue to burn fossil fuels and generate atmospheric carbon pollution, we're pushing the planet harder and harder. The question is, how long before we've pushed *too* hard?

THERE ARE TWO EVENTS from the early paleoclimate record that speak with unusual clarity to these dueling narratives of resilience and fragility. The so-called Faint Young Sun paradox, on the one hand, illustrates the stabilizing (often called negative feedback) mechanisms—forces that have prevented runaway climate behavior during the vast majority of Earth's history. On the other hand, Snowball Earth, an episode (or possibly two) when the blue planet turned into a white snowball, demonstrates the destabilizing (often called positive feedback) mechanisms—vicious cycles that can generate runaway climate change.

The Faint Young Sun Paradox

Earth formed 4.55 billion years ago. The geologic eon that began then and ended 3.8 billion years ago is called the Hadean after Hades, the Greek God of the Underworld. The moniker is well earned. Earth at that time would have been a hellscape, as it continued to be bombarded by planetesimals—moon-sized spheres of rock and dust—that would have wreaked havoc when they struck, vaporizing the oceans, fracturing Earth's surface, and blanketing the sun. One of these planetesimals, a Mars-sized body called Theia—named after the child of the Greek Earth goddess Gaia—hit Earth, and the impact ejected enough material to coalesce and form what is now our moon.

The bombardment ended around 3.8 billion years ago, which marked the beginning of the next geologic eon, the Archean—named after Arche, the muse of origins and beginnings in Greek mythology. Once again, a name well earned. Life on Earth, in the form of one-celled bacteria appropriately termed archaea, appears to have emerged almost immediately when the bombardment stopped. Life—to quote Jeff Goldblum in *Jurassic Park*—seems to find a way. At least when circumstances allow it.

But therein lies the paradox. Back in the early 1970s, Carl Sagan posed the following question: How could Earth have maintained a hospitable climate nearly four billion years ago? At least enough to support those one-celled bacteria? Models of stellar evolution indicate that the Sun back then was thirty percent dimmer than it is today. If you do the calculations, you come up with a frozen planet, one that lacks liquid water—an ingredient that is essential for life as we know it. But don't take my word for it. Let's do the math.

Calculus, integrals, differential equations—these words understandably strike fear into the hearts of most readers. The good news? We don't need any of these things. We just need to plug a few numbers into a plain old algebraic expression. A simple model, known as an energy balance model, calculates Earth's temperature by mathematically balancing the incoming solar energy impinging on Earth's surface (after accounting for the portion that is reflected back to space, mostly by clouds but also ice and other reflective surfaces) with the outgoing heat energy Earth loses to space.[1]

The Stefan-Boltzmann law of physics, sometimes called the black-body radiation law, tells us that all objects radiate energy in proportion to the fourth power of their temperature. For objects at typical room temperature, that radiation is invisible—it lies in the infrared region of the spectrum. But it's there, in the form of radiated heat energy. If you've ever worn night vision goggles, that's what you were seeing. The Stefan-Boltzmann law of course also applies to the Earth, and that has deep implications. The more you warm up the surface of the planet, the more the planet loses heat to space. It's a stabilizing feedback called the Planck feedback, after the great physicist Max Planck, who first worked out the underlying physics. The Planck feedback is the most important stabilizing feature of the climate system.

We end up with a simple expression that says that Earth's surface temperature raised to the fourth power is equal to the average amount of incoming solar heating of Earth's surface, divided by a physical constant. We can now calculate Earth's current surface temperature. The total amount of incoming solar heating today is roughly equivalent to the heat emitted by a typical electric hair dryer on the "medium" setting (1370 watts of power to be precise) for each square meter of Earth's surface. Divide by four to account for the spherical shape of Earth's surface, and multiple by 0.7 because roughly thirty percent of the incoming sunlight is reflected back to space. Plug the numbers into a calculator, hit the square root button twice, and you get the answer: a surface temperature of, wait for it,

255 kelvins. That's −18°C or, in our preferred units, 0°F. A frigid, lifeless planet.

We know that can't be right, because, hello! You're sitting here reading this book. In fact, the average surface temperature of the planet is about 60°F. What went wrong with our calculation? Well, we left out something rather important: the role played by atmospheric greenhouse gases, like carbon dioxide and water vapor. They turn what would otherwise be a frozen, lifeless planet into a habitable one by absorbing some of the heat energy Earth is trying to send out to space, then radiating it back down toward the surface, warming it up. That's the greenhouse effect.

For the purposes of our model, we can measure the greenhouse effect using *emissivity*, which is how effective the Earth's atmosphere is at trapping outgoing heat. Zero emissivity means no greenhouse effect. Earth's greenhouse gases absorb about seventy-seven percent of the outgoing heat energy. If we adjust our calculations to account for this factor, we get the temperature 60°F—the right answer.

So what does this have to do with the Faint Young Sun paradox? Three billion years ago the Sun was only seventy percent as bright as today, so solar heating was diminished. That's like turning the setting on the hair dryer from "medium" to "low," which would put the Earth well below freezing again. We're back to a lifeless planet.

But of course, we know that Earth wasn't lifeless back then, and it wasn't frozen. It had oceans that were teaming with microbes. We have fossil evidence of them dating back at least 3.5 billion years in the form of stromatolites, dome-shaped layered deposits of limestone that were formed by colonies of cyanobacteria. They lived near the ocean surface, employing photosynthesis to capture solar energy. They appear to have been using a primitive anoxic metabolic pathway that causes carbon dioxide and hydrogen sulfide to react, producing energy-rich sugar molecules and, as a by-product, molecular sulfur.

So, what gives? Our calculations suggested a frozen, lifeless planet. The evidence strongly suggests otherwise. Carl Sagan, with coauthor George Mullen, proposed a solution a half century ago that is generally now accepted as basically correct. They hypothesized that a stronger greenhouse effect was at work. The culprit, they speculated, was ammonia gas—a potent greenhouse gas. Ammonia would have been relatively stable in Earth's early, oxygen-free atmosphere because oxidation was absent. But without a protective ozone layer (which would only show up during the oxygenation of our atmosphere a billion years later), ammonia should have been destroyed by high-energy ultraviolet radiation reaching Earth's surface.[2]

Sagan's basic solution to the paradox—a higher greenhouse effect in Earth's early atmosphere—is correct. The greenhouse gases most likely involved, however, were instead carbon dioxide and methane, ironically the two main greenhouse gases that we're releasing into the atmosphere now through fossil fuel extraction, fossil fuel burning, and industrial agriculture.

Carbon dioxide levels would have been high back then for at least two reasons. First, the hotter interior of the young Earth—a remnant of its origins as a ball of molten rock—would have generated more vigorous plate tectonics, and therefore more volcanic outgassing of carbon dioxide into the atmosphere. Second, as we learned earlier, chemical weathering by continental rocks is an important mechanism for drawing down carbon dioxide from the atmosphere. A mostly water-covered Earth would have seen greatly reduced carbon drawdown. The combination of increased production and decreased drawdown translates to higher carbon dioxide levels.

Methane is a different story. As we learned earlier, the first life-forms were probably anaerobic bacteria that got their energy by combining carbon dioxide with hydrogen gas, producing methane as a by-product. In Earth's oxygen-free early atmosphere, there would have been little chemical removal of methane, allowing it to build up more readily in the atmosphere. Even back then, we see that life itself was beginning to play a role in regulating our planetary climate.

So, we have a credible solution to the Faint Young Sun paradox: a super greenhouse of carbon dioxide and methane early in Earth's history. But something else happened subsequently that is equally, if not more, enigmatic. As the Sun gradually grew brighter in the ensuing eons, Earth did not get hotter. Our planet, it seems, has a geologic thermostat. Does this planetary body, like our own human bodies, regulate its own temperature? Let's talk about Gaia.[3]

Gaia

When we examine other planets in our solar system that have an atmosphere and a greenhouse effect, we see that Mars's greenhouse effect is too small, so it's a frozen, barren desert. Venus's is too large, so it's a hellish inferno. Ours is just right. A Goldilocks planet. Is that pure coincidence?

Earth's greenhouse effect slowly weakened as the Sun gradually got brighter, so that the planet's surface temperature remained within hospitable bounds. Could it be that the greenhouse effect somehow adjusted to keep the planet habitable? Sounds absurd, right? Could Earth have a mind or a will of its own? This planetary riddle preoccupied Carl Sagan's onetime life partner, the renowned scientist Lynn Margulis.

As an aside, Margulis, who was briefly a colleague of mine while I was at the University of Massachusetts in the late 1990s, was one of the few scientists in modern history to generate not one but two revolutionary concepts. One of these was *endosymbiosis*, the theory that certain components of cells like mitochondria (responsible for respiration in both plant and animal cells) and chloroplasts (which allow plants to photosynthesize) were originally autonomous single-celled life-forms, like bacteria, that were absorbed into the cells of plants and animals, forming a lasting, symbiotic relationship. Received as controversial and rejected by fifteen scientific journals when Margulis proposed it in 1966, the theory became widely accepted by the early 1980s when DNA studies demonstrated that both mitochondria and chloroplasts have their own distinct DNA, differing from that of their host organisms. (The chloroplast, originally, was a primitive photosynthesizing cyanobacteria of the sort discussed earlier in this chapter.)[4]

Margulis's other groundbreaking innovation—in collaboration with the iconoclastic British scientist James Lovelock—was *Gaia*. Named after the eponymous Greek Earth goddess, the Gaia hypothesis, published in 1974, posits that the Earth system—including life itself—regulates conditions on Earth in a way that keeps the planet in habitable bounds, a sort of thermostat. There are mechanisms that are fundamental to the way our climate system operates, in other words, that make Earth's climate system resilient, at least to a point.[5]

The Gaia hypothesis, like endosymbiosis, was highly controversial from the very beginning, ridiculed, maligned, and misunderstood by Margulis and Lovelock's fellow scientists. It was seen by some as attributing sentience and intentionality to the Earth system, a sort of planetary anthropomorphization. One could argue, in fairness, that naming their hypothesis after a mythical personification didn't help. Representative of the criticisms leveled against the hypothesis, microbiologist and Royal Society Fellow John Postgate complained: "Gaia—the Great Earth Mother! The planetary organism! Am I the only biologist to suffer a nasty twitch, a feeling of unreality, when the media invite me yet again to take it seriously?" By now you're probably suspecting that it's unwise to side with Margulis's critics. And you would be right.[6]

Gaia has at times ironically been embraced by both environmentalists and pave-the-Earthers alike. Environmentalists gravitated to the "Earth Mother" framing that was so prevalent in the environmental movement at the time the article was published in 1974. I was nine years old then, and I remember how pervasive the concept of Mother Nature was in the popular culture of the mid-1970s. "It's not nice to fool Mother Nature" was the tagline of a famous ad I can still recall, for a butter substitute called Chiffon. Mother Nature was the heroine in the now classic 1974 stop-motion animated Christmas special *The Year Without a Santa Claus*. (It was the only thing my elementary school friends and I were talking about in school the next day.) "Look at Mother Nature on the run in the nineteen seventies" go the lyrics of a classic Neil Young song from that time ("After the Gold Rush").

The pave-the-Earthers, on the other hand, liked the Gaian framing of a resilient planet. It seemed to them to depict an Earth system that is largely immune to environmental pollution and other forms of human-generated insult. A rather convenient premise for industrial polluters. But the Gaia hypothesis didn't actually say *any* of these things. It simply proposed that there are processes governing the Earth system that act—through the laws of physics, chemistry, and biology—in such a manner as to generally oppose forces that are pushing the system away from its equilibrium state.

This stabilizing property of our planetary home is not fortuitous. It is a consequence of a weak form of the so-called anthropic principle: of all the planets in all of the solar systems in all of the galaxies of our universe, only planets with Gaian properties are likely to nurture life for billions of years, eventually producing intelligent, self-aware organisms like us that can ponder these matters. (An aside: You could even argue for a stronger form of the anthropic principle: of all possible universes, with all possible values of the fundamental physical constants and versions of the laws of physics, only those universes that produce spiral galaxies and solar systems and laws of chemistry and biology like those that govern our planet will lead to life and, eventually, intelligent beings. It isn't my intent to argue for any one of these competing viewpoints. All that matters, for our purposes, is that there is indeed evidence for Gaian properties on our planet.)

We witness Gaian behavior in Earth's long-term geological carbon cycle. As the Sun gradually grew brighter over time, favoring a warmer planet, the hydrological cycle—the cycle of evaporation and precipitation of water—intensified. More rainfall meant more weathering of rocks and scouring of atmospheric carbon dioxide from the atmosphere, driving down carbon dioxide levels and lowering the greenhouse effect, favoring a cooling of the climate. In this manner, Earth's temperature has remained habitably moderate even as the Sun has become continually brighter.

Gaia is also seen in the increasingly important role that life has played over time in the carbon cycle. Consider the fate of carbon dioxide after it is drawn down by chemical weathering. The burial of carbon initially occurred through the slow, inefficient, inorganic processes discussed in the first chapter. But increasingly over time, as life took hold, more efficient organic burial mechanisms emerged—Gaia at work. The most important among them was the incorporation of carbonates into the skeletons of shell-forming marine organisms like plankton and foraminifera (forams to those in the know) and, later, corals, mollusks, and primitive crustaceans such as the familiar trilobite. These skeletons eventually sink and become cemented together, forming sedimentary rock like limestone on the seafloor. The remains can at times become buried under thick layers of mud, creating oxygen-free (anoxic) conditions. Subject to high pressure and temperature as sediment accumulates and organic matter is buried deeper and deeper beneath the ocean floor, the organic matter is eventually converted into hydrocarbons such as oil and natural gas. Along with coal, they are what we call fossil fuels.

When plate tectonics drive an oceanic crustal plate beneath another plate, limestone rock is subject to tremendous pressure and temperature. The extreme heat combines with the plentiful silicon in Earth's crust to turn the limestone into silicate rocks, releasing carbon dioxide gas. The gas escapes back into the atmosphere through volcanic vents, completing the long-term carbon cycle.

While chemical weathering prevents carbon dioxide levels from getting too high, volcanic outgassing prevents them from getting too low. Perhaps tectonic activity is a crucial ingredient in keeping temperatures within livable bounds. Mars appears to have been tectonically active a couple billion years ago and to have had abundant liquid water on its surface at that time. Maybe it had life. Today it is tectonically dead. It's also frozen and lifeless.

The precise balance between drawdown from weathering and input through volcanic outgassing can shift over time. When plate tectonics are more active, there is more outgassing and atmospheric carbon dioxide levels tend to go up. That gave us the very warm climate of the Triassic and early Jurassic periods around 200 million years ago, when the dinosaurs roamed our planet. As we saw in Chapter 1, however, plate tectonics can also impact the other half of the cycle—the chemical weathering of carbon dioxide. The forced uplift of the Himalayas due to the collision of India with Eurasia fifty million years ago, for example, favored the subsequent drop in carbon dioxide levels through a strengthening of the Asian summer monsoon, greater rainfall, and more weathering.

Despite the gradual rise and fall of carbon dioxide over millions and billions of years, Earth's climate has remained largely within livable bounds, staying far away from both the infernal conditions of Venus and the frozen, barren conditions of Mars today. We've seen now how life, over time, has played an increasingly important role in the stabilizing mechanisms of Gaia. Threatened by the chilling prospects of a dim Sun, Earth responded with methanogenic bacteria to warm the planet up. Threatened instead with an increasingly *hot* Sun, Earth responded by burying carbonaceous life at the ocean bottom, drawing down carbon dioxide levels. Life, once again, found a way.

As early as two billion years ago, evolution figured out a new, considerably more efficient photosynthetic pathway for cyanobacteria to employ, by splitting water molecules, the supply of which, unlike hydrogen or hydrogen sulfide, was essentially unlimited. A by-product of this pathway is oxygen, so oxygen levels in the atmosphere skyrocketed. Oxygen allowed for a whole new metabolic pathway for life—respiration. But it was also toxic to many anaerobic bacteria, which experienced a mass die-off. As with the climate's effect on humans, with bacteria there were winners and losers.

By 600 million years ago, so much oxygen had accumulated in the atmosphere that it was literally piling up to the stratosphere where it encountered high-energy ultraviolet (UV) radiation. UV rays split up oxygen molecules, combining the free oxygen atom with an oxygen molecule to make ozone. So, Earth now had an ozone layer to absorb damaging UV radiation before it reaches the surface. That made it safe for life to come up to the surface of the ocean, and to crawl onto the land.

By 500 million years ago we had land-based plants and by 400 million years ago the first terrestrial animals (the oldest preserved specimen is a millipede-like creature). They both aided in the sequestration of atmospheric carbon. When plants and animals die, their organic remains find their way into soils or muddy swamp bottoms. The geological process of sedimentation works its magic. They become buried in layers of rock and dirt. Over millions of years, heat and pressure from Earth's crust decomposes these organisms into what we know as coal. It took Gaia more than a hundred million years to bury this carbon. Today, we're unburying it and putting it back into the atmosphere over a timeframe of a hundred years, *a million times faster*.

Around 400 million years ago we also saw the rise of vascular plants—plants with roots, stems, and leaves. Robert Berner, a former professor of mine in the Yale Geology and Geophysics Department, showed that the evolution of vascular plants allowed life to exert even greater control on the Earth system by recycling water and by producing acids that accelerate the process of chemical weathering. Gaia, if you like, has been growing ever more influential over time and, increasingly, her favored tool is life itself. The question is, can Gaia survive us?[7]

Daisyworld

James Lovelock was an unconventional scientist, known for his brilliant, if often controversial, ideas (like the Gaia hypothesis he formulated in collaboration with Lynn Margulis). Having received a Ph.D. in medicine in 1948, he went on to do biological and astronomical research at Yale and Harvard, among other institutions. He was a prolific author of more than a dozen books and numerous scientific publications and received a number of prestigious scientific prizes and awards, including election to the Royal Society in 1974.[8]

In the mid-1970s Lovelock was an adviser to the NASA Viking mission to Mars. He argued that prospects for life on other planets could be assessed by measuring the composition of their atmospheres by telescope. Drawing on the lessons learned from the origins of life on Earth, he suggested that an atmosphere rich in oxygen and methane would likely signal the presence of living organisms. But a planet with an atmosphere like Mars's today, composed largely of carbon dioxide and nitrogen, would be unlikely to harbor life. The orbiters and landers of the Viking mission returned fascinating images and detailed information about the red planet. Among other things, they found geological evidence of deep river valleys that were likely carved by torrents of water at some point in Mars's distant past, a tantalizing clue that conditions might have been far more conducive to life early on in the planet's history. But the Viking mission indeed struck out when it came to finding evidence of life today.

Lovelock passed away in 2022 at the ripe old age of 103. It would seem he knew a thing or two about longevity. And he spoke with particular authority when it came to Gaia and her more-than-three-billion-year life span. Troubled, however, by the anthropomorphization and misinterpretations of the Gaia hypothesis by the public and even other scientists, Lovelock set out in the mid-1980s to construct a model that would demonstrate how Gaian behavior can arise purely from the laws of physics, chemistry, and biology. No need for sentience, will, or desire.

We can all agree that daisies don't think. They don't plan. They have no ambitions, worries, or agenda. They are slaves to their environment. What if I described a planet with nothing but daisies where these flowers, without motivation or intention, driven simply by laws of physics and biology, acted collectively to keep planetary temperature within livable bounds. Would you believe me?

We're talking about Lovelock's hypothetical "Daisyworld." It is just like Earth in most respects—the same distance from its sun, receiving similar amounts of sunlight. But it has no oceans. It has only land—with a layer of soil. There is a single life-form that can potentially grow in the topsoil: white daisies. While the soil absorbs a fair amount of sunlight, reflecting only twenty percent of it, the white daisies are highly reflective, reflecting ninety percent of the incoming sunlight. So, a planet covered in daisies will have a higher reflectivity, and therefore a cooler temperature, than one covered in bare soil.[9]

Daisies, on our hypothetical planet, prefer conditions that are not too hot and not too cold. They can tolerate temperatures in the range of 54°F to 104°F (12°C to 40°C), but the optimal temperature, where they become most widespread, is 79°F (26°C). Outside the range of temperatures they can tolerate, the daisies die.

So, we have Daisyworld, an interactive system with feedback mechanisms. Daisy coverage impacts the temperature of the planet by changing the reflectivity of the surface and the amount of solar radiation reflected back to space. But temperature, in turn, impacts the distribution of daisies. We call such systems *nonlinear*: they can exhibit large changes when subject to small perturbations. This can lead to complex and sometimes surprising behavior.

Just as Earth has a planetary history, so does our imaginary Daisyworld have its own story. It was once completely barren. We might have called it Barrenworld back then. The bare soil absorbed most (eighty percent) of the incoming sunlight, but there was too little of it to warm the planet beyond the freezing point. Its sun was only sixty percent as bright as our own Sun is today. It was too cold for life. The sun slowly brightened, however, eventually reaching seventy percent of the brightness of our Sun. That warmed up the planet to a mild temperature of 54°F (12°C). Alien life quickly evolved under these favorable conditions—though it was just a single species, daisies (which happen to be identical to Earth daisies in a remarkable coincidence of convergent interplanetary evolution). Daisyworld's sun continued to brighten, and temperatures started to warm a bit, but as the number of daisies increased, they reflected back more and more of the incoming sunlight, limiting the warming. The daisies were thriving as the sun reached the level of our own current Sun. In fact, nearly three quarters of the planet was covered by daisies. They were now reflecting more than seventy percent of the incoming sunlight, keeping the planet's temperature comfortably moderate, within livable bounds. Life was providing a key stabilizing feedback.

But Daisyworld's sun continued to get brighter. Eventually, it climbed to 110 percent of our own Sun's brightness. The daisies reached their maximum extent, covering eighty percent of the planet's surface, and they were reflecting most (seventy-five percent) of the sunlight back to space. But that wasn't enough to offset the warming effect of the planet's ever-brightening sun. As that sun continued to brighten further, the daisies began to die off. That meant they were reflecting less sunlight, so the planet absorbed more and warmed even more. And more daisies died. The situation rapidly spiraled out of control. Soon, the daisies were entirely gone, and the reflectivity of the planet dropped to twenty percent, the baseline level for barren soil, with the planet now absorbing eighty percent of the incoming sunlight. The temperature quickly spiked to an infernal 140°F (60°C). The stabilizing feedback that had for so long helped preserve life on Daisyworld had disappeared. It had been instead replaced by a destabilizing feedback.

Daisyworld was gone. In its place was once again the lifeless Barrenworld from whence it came. Some astronauts from Earth arrived in an expedition to the planet. They drilled some sediment cores that showed that daisies had once thrived on the planet, back when the planet was cooler, in the 70s°F (low/mid 20s°C). In an attempt to colonize life once again on the planet, they engaged in a massive geoengineering effort, shooting prodigious amounts of reflective particles into the planet's upper atmosphere to block out incoming sunlight and cool the planet down. They were able to reduce the amount of sunlight reaching the surface to ninety percent of our own Sun's brightness, a level at which they knew the daisies had once flourished on Daisyworld. But that only brought the planet's temperature down to a still rather hot 105°F (41°C) as the barren desert soil continued to absorb eighty percent of the incident solar heating in the absence of daisies. The Earth daisies that crew botanist Mark Whatley attempted to plant in the soil didn't take. They died off immediately in the oppressive heat. The astronaut crew gave up and returned to Earth unsuccessful in their efforts to revive Daisyworld.

What lessons can we learn from this story? Lovelock's Daisyworld is obviously a gross simplification of the actual Earth system. But all useful models are simplifications. And it is the simplicity that often reveals basic truths. In this case, the truth is that life really can act in a way that benefits itself by helping keep its host planet within livable

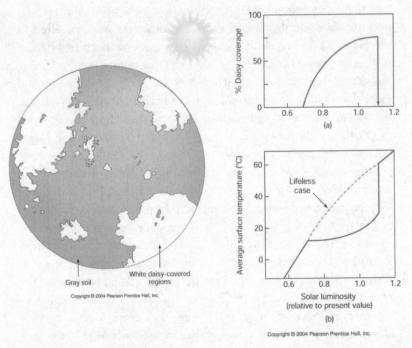

Figure 6. An illustration of Daisyworld. The solid curve represents Daisyworld, and the dashed curve represents Barrenworld.

bounds. There is no feature of the Daisyworld model that involves the desire of daisies or the will of Gaia. There are no sentient actors. There's just physics and biology. The potential fallacy on our part is to anthropomorphize the behavior of the system and interpret it as intentionality. Indeed, the tendency of the human brain to find motive, intention, and patterns where they don't actually exist isn't a coincidence. Evolution favored it—our ancestors who wrongly thought they saw a lion in the dark of night were more likely to survive than those who didn't see the lion that really was there. Perhaps some folks are doing something similar when they ascribe intentionality to the Earth system.

This simple conceptual model demonstrates the resilience of the climate, but only over a restricted temperature range. The daisies behave in a way that keeps the planet far more habitable than it would otherwise be. Life finds its way, but only up to a point—a *tipping point* where a dramatic warming spike occurs and life disappears in an instant. Daisyworld, among other things, is a conceptual model for climate-induced collapse.

Let me comment on one other important implication of this simple model for the Earth system. I'm often challenged by people (on social media in particular) when I refer to the threat to our *planet* posed by human-caused environmental degradation. Their retort? "The *planet* itself will be just fine." Yes, it's true that there will still be a huge spherical object with the dimensions of Earth and the same proximity to our star if we extinguish life on this planet. But the *planet*, as we know it, with its climate and other characteristics, has been fundamentally shaped by life itself. Just as Daisyworld becomes Barrenworld if you kill off the daisies, our planet will no longer be the thriving living planet we call Earth if we continue headlong down the unsustainable path we're currently on.

Snowball Earth

Earth has been subject to the stabilizing Gaian influences for billions of years, distinguishing it from its hotter, more Sun-proximal neighbor Venus, which underwent a runaway greenhouse effect more than two billion years ago. But it appears that there *have* been episodes (one or *possibly* two) of runaway behavior in Earth's climate. Oddly enough, they ran in the opposite direction—not a runaway greenhouse but a *runaway icehouse*.

During Snowball Earth episodes, the entire surface of the planet, from pole to equator, was covered in a shell of ice. The first such episode appears to have occurred a little more than two billion years ago during the early part of the Proterozoic, the geological eon that followed the Archean. It's a particularly long eon, extending from 2.5 billion years ago to 541 million years ago. The early part is known as the Paleoproterozoic. The late part is called the Neoproterozoic. There *might* have been a Snowball Earth episode then, too. We'll get to that in a minute.

The Sun is estimated to have been only about eighty percent as bright as today during the Paleoproterozoic, and the planet was consequently more prone to glaciation. We've known about the existence of episodes of extensive continental ice covering substantial regions of the planet—called the Huronian glaciations after Lake Huron, near which evidence of it was first discovered—for a long time. But it was only in the late 1990s that we had enough evidence amassed to suggest that the entire planet, from the poles to the tropics, was covered in ice. Deep into the tropical regions were striated rocks that showed scrape marks glaciers would have made as they moved over them. There were dropstones, which are rocks that were picked up by glaciers and dropped in the ocean. And there were tillites—rocks made up of the debris, or till, left behind when a glacier moves past rock—that appear to have come from the equatorial region, close to sea level. All of this empirical evidence implies that the ice was pervasive, covering both land and ocean, all the way down to the equator. Cap that off with evidence of cap carbonates—thick layers of carbonate sitting on top of glacial deposits (we'll see why that's important a bit later)—and the collective evidence strongly suggests a Snowball Earth episode.[10]

The Paleoproterozoic Snowball Earth event closely coincided, almost certainly not coincidentally, with what is known as the Great Oxidation Event. Cyanobacteria, as we learned earlier, had by this time developed a far more efficient photosynthetic pathway using water and producing oxygen. Oxygen levels spiked soon thereafter as a

Figure 7. Geological evidence of glaciation including (top) a tillite from 2.4 billion years ago, (middle) glacial striations from 0.65 billion years ago, and (bottom) a dropstone from 2.4 billion years ago.

result. The rapidly rising oxygen cannibalized atmospheric methane, a potent greenhouse gas we know to have been plentiful in Earth's early atmosphere and which played an important role in resolving the Faint Young Sun paradox. As methane plummeted, the greenhouse effect decreased, cooling the planet. That cooling led to the growth of ice sheets, reflecting more sunlight back to space, reducing the amount of water vapor (a potent greenhouse gas) in the atmosphere, cooling the planet further and forming more ice, until the ice reached the equator.[11]

Our planet, simply put, seems to have turned into a giant frozen snowball. Life acted in a distinctly *non*-Gaian manner during this episode. By spiking the atmosphere with oxygen, microbes threatened all life on the planet. Reckless, you might say, if they were capable of motive or thought. They're not. But we are. And we, too, are currently engaged in actions—foremost among them the profligate burning of fossil fuels—that increasingly threaten life on the entire planet. Yet we don't have the same excuse as the microbes.

Gaia advocates might dismiss this as the exception that proves the rule. But at least one paleontologist, Peter Ward, argues that episodes such as the Great Oxidation Event fundamentally contradict the Gaia hypothesis. In *The Medea Hypothesis*, Ward invokes Medea, a figure of Greek mythology who killed her own children, in arguing that life, rather than acting in ways that are beneficial to life, actually threatens its own existence. I come down somewhere in the middle. Life, in many cases, does appear to act in a Gaian manner. But there were certainly instances in Earth's history—like the Great Oxidation Event—when Medea reared her head.[12]

Speaking of life, one might well ask how it is that it survived a Snowball Earth event. The extremophiles, which include various types of archaebacteria, that inhabit the hydrothermal vents at the ocean floor were insulated from this event. Today, microbes can be found in other extreme subsurface environments like Lake Vostok, a freshwater lake that lies more than two miles beneath the surface of the Antarctic Ice Sheet. It's easy to imagine similar microbes having lived beneath even thick layers of sea ice.

We know that photosynthetic cyanobacteria also made it through. There are a number of other potential so-called refugia (places of refuge) where they could have survived. At the equator, sea ice might may have been no thicker than thirty feet or so. That's thin enough to allow for substantial amounts of sunlight to penetrate and photosynthesis to occur. Volcanoes would have continued to spew material into the atmosphere, darkening the surface ice with ash in places, lowering its reflectivity and increasing absorption of sunlight. It's conceivable that the tropics would have been populated by small pools of water on the ice surface where photosynthesizing microbes could exist.[13]

Returning to the matter of Snowball Earth itself, another even more widely publicized putative snowball episode, which graced the cover of *Scientific American*, is argued to have occurred toward the end of the Proterozoic (the Neoproterozoic), roughly 800 to 550 million years ago. Significantly more recent than the Paleoproterozoic event, the Sun would not have been as dim (around ninety percent as bright as today), posing a greater challenge for tropical glaciation. Paul Hoffman of Harvard University and coauthors argued back in the late 1990s that glacial deposits in Namibia dating to this time interval show evidence of a total shutdown in ocean biological activity—a scenario consistent with an extended Snowball Earth event. What might have precipitated this event? Proponents argue that the rise of multicellular life, including the colonization of land, expansion of terrestrial lichen and fungus-rich ecosystems, and development of organic soils, could have caused a rapid rise in oxygen concentrations similar to the Paleoproterozoic Great Oxidation Event. But my former Penn State colleague Lee Kump argues instead that consumption of oxygen by aerobic microbes in newly plentiful organic soils would have countered any tendency for rising oxygen levels. Other studies suggest that the glaciations themselves might have actually caused the diversification of life and development of multicellular organisms at this time.[14]

There is indeed a healthy ongoing debate about whether or not there actually was a Neoproterozoic Snowball Earth episode. Some argue that geochemical evidence is inconsistent with the slowdown in chemical weathering that would be expected on a frozen planet. Others argue that continental uplift may have raised rocks that have been interpreted as near sea level to high enough elevations to induce glaciation. The fact that there were multiple distinct glacial intervals that were interrupted by warm periods calls into question the idea that there was a single, persistent snowball evident. It instead suggests cyclical alterations between glacial and interglacial periods that are more reminiscent of the Pleistocene ice ages. Other researchers have argued that many glacial deposits from that period lack the putative cap carbonates and that there are non-glacial explanations for the apparent cap carbonates that have been found, involving instead the rapid and widespread release of methane. Yet others argue that the assumptions behind paleolatitude estimates, which are based on ancient geomagnetic data inferred as indicating proximity to the pole, may be flawed and that glaciation might not have occurred at low latitudes. In an article in 2000 published in the leading journal *Nature*, paleoclimate modeler William Hyde and coauthors performed computer simulations, driving the models with greenhouse gas concentrations and solar brightness appropriate for that time period. Their simulations indeed allowed for a global belt of ice-free open water centered at the equator. Such an alternative "Slushball" Earth would have provided equatorial refugia for life, as well as a continuation of the hydrologic cycle. The authors argue that their alternative scenario better explains the unusual configuration of tropical carbonates characteristic of the Neoproterozoic sedimentary record. Neoproterozoic Snowball Earth proponents Dan Schrag and Paul Hoffman of Harvard have in turn argued that the available geochemical evidence is inconsistent with a band of open equatorial ocean.[15]

The details and even existence of Snowball Earth episodes is still hotly (forgive the pun) debated, in good faith, based on evidence and logical arguments by scientists on both sides. That's the way science is supposed to work. As an author of a book about climate history, you're called upon to referee that debate as best you can. In my assessment, based on all of the evidence to date, there very likely was an early, Paleoproterozoic Snowball Earth episode. I'm somewhat less convinced there was a later, Neoproterozoic Snowball Earth episode. In any case, it isn't necessary for the larger point that Earth, at least once, seems to have found itself in a snowball state.

The White Earth Solution

How did Earth get out of this predicament once it was in it? To answer that question, we'll need to talk once again about feedbacks, both stabilizing and destabilizing. With Snowball Earth we see another important destabilizing feedback, the so-called ice albedo feedback (*albedo* is a fancy scientific term for reflectivity). That's what gets us stuck in a snowball state in the first place. The albedo is the percent of incoming solar energy that an object reflects. Ice, like white daisies, reflects much (sixty to eighty percent, depending on how old/dirty the ice is) of the incoming solar energy. We can contrast that with the far lower twenty percent reflectivity of soil. If the entire planet becomes covered in ice, it reflects most of the incoming sunlight, keeping it cold and ice-covered.

Back in the late 1960s, Russian climate scientist Mikhail Budyko and American climate scientist William Sellers independently, and essentially simultaneously, created a simple model of Earth's climate that led to a curious conclusion: under the right conditions, our planet could become covered in ice. The cleverly named Budyko-Sellers model is a bit more complicated than the very simple model we encountered at the beginning of the chapter that treated Earth as a mathematical point. In that case, there was no latitude or longitude, just a single temperature to represent the entire planet.

The Budyko-Sellers model introduces latitude as a variable. That adds complexity, but not so much complexity that I can't assign it as an exercise to my undergraduate students. Here's how it works. We now divide Earth into different latitude bands. There's a polar band (70–90 degrees latitude), a subpolar band (50–70°), a mid-latitude band (30–50°), and a tropical band (10–30°) in each hemisphere, as well as an equatorial band (10°S–10°N). In each band there is incoming solar heating (the most at the equator, and the least at the poles) and there is loss of heat energy to space. But here's the rub: there's now a surplus of energy at the equator (more incoming solar energy than outgoing heat energy) and a deficit at the poles (more outgoing heat energy than incoming solar energy). That means that there has to be a flow of energy from low to high latitudes to prevent the poles from getting colder and colder and the equator from getting hotter and hotter. To deal with that, the model includes a "heat flux" term that carries heat poleward. It's a stand-in for the role that ocean currents and atmospheric circulations play in transporting heat from the equator to the pole.[16]

Here's where things get interesting: The albedo (reflectivity) can vary depending on whether a given latitude zone is cold enough to have ice or not. Ice can form first at high latitudes, which cool down as a result, transferring that cold to lower latitudes, which then cool down and form ice, and so on. Through this process, it's possible to build up an ice sheet. What Budyko showed was that not only could you build up an ice sheet, but you could also grow it all the way down to the equator. And there we have it: Snowball Earth!

The Budyko-Sellers model exhibits unusual behavior that is sometimes present in nonlinear systems. It's known as a *hysteresis loop*. If you've had the unpleasant experience of being unable to set the temperature dial at a comfortable level in your hotel shower because the dial behaves differently turning it up and turning it down (making it hard to find the "just right" temperature), then you've suffered an unfortunate, but hardly life-ending, encounter with the phenomenon.

We already saw an example of hysteresis in Daisyworld. Consider what happens when you start with a sun that's dim (say, eighty percent of our Sun's current brightness), but the temperature is warm enough, around 60°F (16°C), for daisies. As the solar brightness increases further, we eventually reach the point (roughly 115 percent of current brightness) when it becomes too hot for them. Now let us suppose that it increases further, to 120 percent, and stays that way for millions of years. It's safe to declare that scenario game over for daisies. There is no more Daisyworld. It's Barrenworld now. What if, for some reason, the sun starts to grow dimmer and temperatures begin to cool down. We reach 115 percent current sunlight, then 100 percent and all the way back to eighty percent. But there are no daisies anymore, so we follow the hotter Barrenworld curve in Figure 6. The surface temperature isn't a pleasant 60°F (16°C) now but a scorching 90°F (32°C). We turned the livable planet of Daisyworld into a hot, lifeless, barren wasteland, and there's no going back now. Add that to our growing list of cautionary tales.

The phenomenon of hysteresis is directly relevant to the behavior of Earth's climate system today. Even if we warm the planet up enough to melt the ice sheets, there's a chance we could cool the climate back down over the next century (like through artificial carbon capture technology that might become viable in the future). But it's not as if the ice sheets will return. They're done. It would take millions of years to bring them back. A similar thing holds with the great ocean conveyor. If that circulation pattern collapses due to warming and we cool the climate back down, that circulation pattern doesn't come back. These systems are nonlinear, and they exhibit tipping point behavior—behavior that is irreversible on human timescales. And because of the nature of scientific uncertainty, we don't know precisely how close we are to these tipping points. That fact should give us pause as we continue to recklessly warm our planet with carbon pollution.

Returning now to the matter at hand, the Budyko-Sellers model exhibits hysteresis when you turn the heating up or down. That change in heating could result from altering the brightness of the sun, the concentration of greenhouse gases, or some combination of the two. Let's assume that we start with heating that is thirty-five percent greater than today, where the model predicts an ice-free planet. If you start there and slowly decrease the heating, no ice forms until you reduce heating to ten percent below today's level. Ice then starts to form near the poles. As the heating is decreased further, the ice quickly spreads to lower latitudes, and by the time you reach eighty-five percent of today, the entire planet freezes over.

On the other hand, suppose that we start with heating that is twenty percent lower than (eighty percent of) today. That's equivalent to the reduction in solar heating (assuming no change in greenhouse gases) at the time of the Paleoproterozoic Snowball Earth event. The model in this case predicts an ice-covered planet. As the heating is increased, the planet remains frozen and ice-covered until the heating reaches a level thirty percent greater than (130 percent of) today. The ice then melts at the equator, and the melting quickly spreads to higher latitudes. Finally when you reach heating of thirty-five percent greater than today, the planet becomes ice-free.

Whether or not we have an ice-covered planet is thus seen to depend on where we started. At current levels of heating, there are two possible solutions: ice-free (which is roughly the case for Earth today) and ice-covered. Budyko called the latter state the "white Earth solution." It would later be named Snowball Earth by Caltech geophysicist Joseph Kirschvink. Systems that exhibit this sort of behavior are said to have multiple steady states. Which state Earth ends up in given today's solar heating and greenhouse gas levels depends on its history. Did we start with a frozen planet and warm it up or with a hothouse planet and cool it down?[17]

A large shock can cause the system to literally "jump" from its current state to another one. In the case of the great ocean conveyor, that shock could take the form of a rapid meltwater release like the one that caused the conveyor to collapse during the Younger Dryas

event we talked about earlier. It could also take the form of the melting of ice today from human-caused greenhouse warming. With the
Paleoproterozoic Snowball Earth, the shock likely took the form of a
rapidly decreasing greenhouse effect as swiftly spiking oxygen scav-
enged the planet-warming methane, plunging Earth into an ice-covered
state.

The problem is that it's very hard to get out of a Snowball Earth
state once you're in one. It is this fact that puzzled Budyko and even
led him to be (wrongly—it turns out!) skeptical of his own findings.
But there was a way out. It just involved science that hadn't yet been
done, including Sagan and Mullen's later Faint Young Sun work,
which established that the Sun was substantially fainter, and Earth
more susceptible to glaciation, early in its history. More importantly,
though, Budyko was focused on the *physics* but not enough on the
geology. Both are important here.[18]

Building on earlier work, other Earth scientists would establish an escape route for a frozen Earth in the early 1990s. Joseph
Kirschvink, the scientist who coined the term *Snowball Earth*, realized that Earth's carbon cycle was the restoring force that could
lead Earth out of a snowball state. Carbon dioxide levels continued
to increase as volcanic eruptions spewed the gas into the air because
the frigid, dry atmosphere had no way to scrub itself of carbon dioxide as the hydrological cycle and chemical weathering came to a
halt. Carbon dioxide eventually climbed to such high concentrations
(perhaps as high as 90,000 ppm—that's more than 200 times higher
than today) that a super greenhouse effect could take hold, spontaneously thawing the icy shell that had formed around the planet in
a catastrophic meltdown. The ice albedo feedback now acted in the
opposite direction from when the Snowball Earth formed, leading to
an accelerating cycle of rapid ice loss, which fed warming, which fed
more ice loss.

With the ocean surface once again exposed, the hydrological cycle
kicked back in, along with chemical weathering. That steadily drew
down the huge mass of atmospheric carbon dioxide, highlighting
now the importance of stabilizing feedbacks. The carbon dioxide was
absorbed into streams and rivers, and ran off to the oceans, where it
was taken up by calcareous plankton, which died and sank, forming

a layer of calcium carbonate, or limestone, at the sea bottom. The resulting stratigraphy identified by geologists millions of years later would show a layer of snowball-era glacial sediment capped by a thick layer of carbonate. And that's the hypothetical origin of the cap carbonates.

Lessons Learned

The Faint Young Sun episode demonstrates the resilience of our planet, to an extent. Earth maintained a habitable climate nearly four billion years ago despite the challenge of a dimmer Sun. As the Sun gradually grew stronger over the ensuing billions of years, the greenhouse effect grew weaker, in such a way that Earth's temperature remained within favorable bounds (most of the time). And as we saw with Daisyworld, life—through its increasingly important role in the global carbon cycle—helped keep conditions suited to itself. We see the hand of Gaia.

Yet, we can also see the hand of Medea if we look for it. Earth's climate spun out of control on at least one occasion—the Paleoproterozoic snowball event. And life itself seemed to play the critical role in that episode: a spike in oxygen due to the rise of photosynthesizing bacteria cannibalized the greenhouse gas methane, cooling the planet. A snowball tailspin ensued, thanks to the destabilizing ice albedo feedback.

We eventually emerged from this Snowball Earth event millions of years later thanks to a countervailing *stabilizing* feedback in the geological carbon cycle. Volcanoes continued to spew carbon dioxide into the atmosphere, but the frozen, dry atmosphere was unable to scavenge it through the normal means of chemical weathering, and so much carbon dioxide accumulated in the atmosphere that the greenhouse effect eventually overpowered the ice and melted it spontaneously. So Gaia prevailed—she made out just fine in the end. But the numerous organisms that went extinct during this catastrophic event? Not so much. The slow, stabilizing carbon cycle feedbacks were of little consolation to them. Thus, we play the role of Medea today, defying Gaia's best efforts to keep our planet habitable. Once again, she may indeed make out just fine in the end. But will we?

The twin tales of the Faint Young Sun and Snowball Earth are the angel and devil, sitting on our shoulders, tugging our climate alternatively toward stability and chaos. For us to stay in this envelope of habitability, the angel must win out. And for that to happen, we must listen to our *better angels*. Yes, we are beneficiaries of stabilizing feedback mechanisms that have helped maintain a favorable climate for us and our civilization. But we could easily become victims of aggravating feedback mechanisms that amplify the warming and destabilize our climate if we continue on our current course of fossil fuel burning.

We know that climate feedback mechanisms—specifically, the destabilizing ice albedo feedback—could lead to runaway glaciation. But Mikhail Budyko recognized the opposite possibility of runaway ice melt when that feedback acts in the opposite direction, where warming begets ice melt, which begets more warming. Indeed, he demonstrated the key role played by that feedback process in modern human-caused warming. Also among Budyko's predictions, made back in 1972, was that fossil fuel burning would melt about half the Arctic ice cap by 2020. The prediction was spot on. He also predicted that the ice albedo feedback would lead to an ice-free Arctic by 2050. That's where we're currently headed in the absence of concerted action.[19]

Climate models tell us such a scenario can still be avoided if we take dramatic measures to reduce carbon emissions. But they must be brought down by fifty percent over the next decade and to zero by mid-century to keep warming below 1.5°C (2.7°F) and preserve at least some of that Arctic ice. If we go much beyond that, we risk losing that ice cap—and potentially much more: large parts of the Greenland and West Antarctic Ice Sheets, leading to massive sea level rise and coastal inundation, not to mention withering heat waves and droughts, unprecedented floods, and deadly superstorms. Our benevolent but fragile moment could transition to a malevolent future. Terrifyingly, the past offers us a window into what that future might look like. It's called the *Great Dying*.[20]

3

The Great Dying Wasn't So Great

With our evolved busy hands and our evolved busy brains,
in an extraordinarily short period of time we've managed
to alter the earth with such geologic-forcing effects that
we ourselves are forces of nature. Climate change, ocean
acidification, the sixth mass extinction of species.

—KATE BERNHEIMER, *XO Orpheus*

We now know that amplifying mechanisms can freeze the
planet, as was the case with Snowball Earth. But these same
mechanisms can also lead to inhospitably hot climates when enough
carbon dioxide enters the atmosphere. Arguably the greatest extinction event of all time—called the Great Dying—appears to have
resulted, at least in part, from a massive heat-inducing release of carbon into the atmosphere 250 million years ago. Is this ancient event
a possible analog for a sixth, human-caused, climate-change-driven
mass extinction today? In answering this question, we will at times
work our way through some details of the science, but the payoff is
that we will see not just that scientists are able to unravel such mys-
teries but how they do it.

Dragonflies and Dinosaurs

When last we left off, it was late in the Proterozoic eon, around 550 million BP. Earth had thawed out from a series of major glaciations,

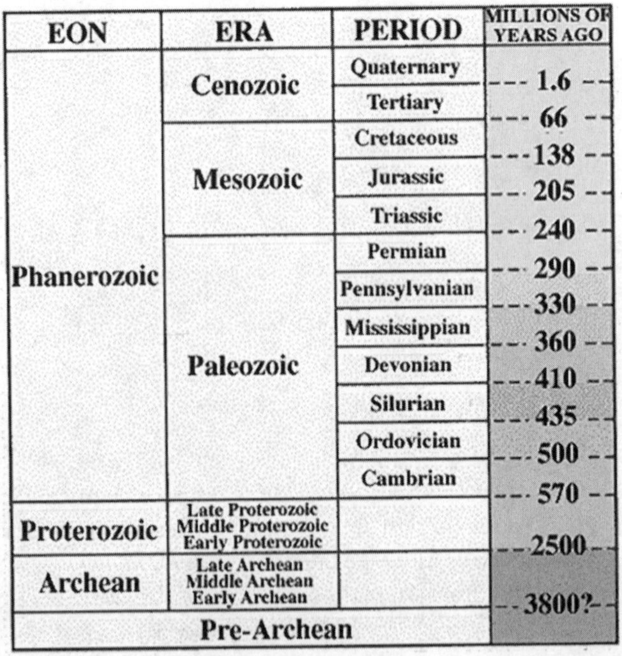

EON	ERA	PERIOD	MILLIONS OF YEARS AGO
Phanerozoic	Cenozoic	Quaternary	--- 1.6 --
		Tertiary	--- 66 --
	Mesozoic	Cretaceous	---138 --
		Jurassic	-- 205 --
		Triassic	-- 240 --
	Paleozoic	Permian	-- 290 --
		Pennsylvanian	---330 --
		Mississippian	-- 360 --
		Devonian	---410 --
		Silurian	-- 435 --
		Ordovician	--- 500 --
		Cambrian	--- 570 --
Proterozoic	Late Proterozoic Middle Proterozoic Early Proterozoic		-- 2500 --
Archean	Late Archean Middle Archean Early Archean		--3800?--
Pre-Archean			

Figure 8. Chronology of the Paleozoic era, in the context of neighboring eras and eons.

perhaps even global snowball conditions. The end of the Proterozoic marked the beginning of a brand new era—the Paleozoic, which extended from around 540 million BP to 251 million BP.

The first period of the Paleozoic—the Cambrian—saw a remarkable explosion in the diversity of life, known, appropriately, as the Cambrian explosion. Most of the life that exists today emerged during the first ten million years of that period, including the first complex multicellular life and familiar groups such as mollusks and crustaceans. Among the reasons for this remarkable diversification was a sustained rise in oxygen from photosynthetic life. Higher levels of oxygen allowed for more diverse, multicellular organisms because they require oxygen in high enough concentrations that it can reach interior cells. The stratospheric ozone layer, which had developed during the Neoproterozoic era, protected animals from the Sun's damaging ultraviolet rays and helped populate the land. Some researchers even argue for a possible "bottleneck" effect, where the few life-forms that survived the Neoproterozoic ice ages (Snowball Earth or not) were able to rapidly fill emerging niches as Earth thawed.

A major glacial event occurred at the end of the following period of the Paleozoic, the Ordovician, around 450 million years ago, as chemical weathering outpaced the volcanic emissions of gas and atmospheric CO_2 levels dropped. The resulting cooling caused a buildup in ice mass on the large South Pole–centered supercontinent of Gondwana. Sea levels dropped. Much of the coastal habitat that had been home to primitive mollusks and crustaceans disappeared. Some of the creatures scraped by, but about half of all existing genuses perished. Much as we can only wonder today what knowledge was lost in the ransacking of the Library of Alexandria, we can also ponder what sort of magnificent creatures born of the Cambrian explosion were lost. Welcome to the first of the widely recognized global mass extinction events. It will hardly be the last we encounter.[1]

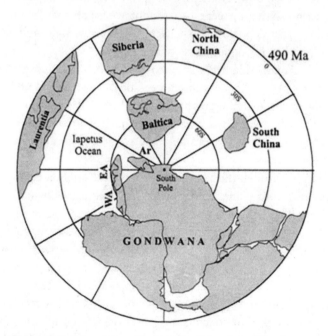

Figure 9. Continental position and configuration of Gondwana during the Ordovician.

The most well-known extinction event ended the reign of the dinosaurs roughly sixty-six million years ago. But the *deadliest* extinction event took place at the end of the Permian period, roughly 250 million years ago. It is referred to in the scientific community as the Permian-Triassic (or P-T for short) extinction, but because an estimated ninety percent of all Permian species disappeared from the face of the planet, it has earned a nickname: the Great Dying. Marine organisms were hit especially hard, with ninety-six percent of species perishing. Gone were the trilobites so familiar to amateur fossil collectors everywhere—primitive arthropods that were the distant ancestors of the modern horseshoe crab. Having survived the earlier Ordovician extinction event, their own nearly 300-million-year moment had come to an end.

Not only were the vast majority of marine invertebrates gone, but so were the earliest fish species. On land, more than two thirds of amphibian and reptile species and nearly one third of insect species were wiped out. Another iconic species, a giant dragonfly called *Meganeuropsis* with a nearly three-foot wingspan that is often included in artist depictions of the Carboniferous period—and to this day still haunts my nightmares—was now gone.

The P-T extinction event wiped out many of the groups that had dominated life on land, freeing up ecological niches to be filled by new organisms, including reptiles such as crocodiles and the earliest dinosaurs. Once again, there were both winners and losers. Who won and who lost, in this case, came down to geology and geochemical weathering cycles.

Midway through the Paleozoic, around 420 million BP, we saw the emergence of plants with roots, stems, and leaves, which as we now know helped accelerate chemical weathering by producing acids that dissolve rock and by helping cycle water from the soil back into the atmosphere. This may have led to a slow, steady decrease in atmospheric CO_2 levels through the late Paleozoic. The spread of these vascular plants, however, also led to a new source of organic matter that could be buried on land or carried off in rivers for ocean burial. Increased burial of organic matter causes rising atmospheric oxygen levels because that organic matter is the product of photosynthesis, which splits up oxygen and carbon atoms. The carbon, once buried, is no longer available to cannibalize the liberated oxygen. In the Paleozoic, oxygen concentrations climbed as high as thirty-five percent (almost twice the current concentration of twenty-one percent).[2]

Those high oxygen levels favored synapsids, creatures with a high metabolism, featuring a single hole in each side of their skull that led to improved jaw function. They were part of a diverse group of four-legged terrestrial animals, including carnivores, insectivores, and herbivores, that first arose in the late Carboniferous and would evolve into the group we today know as mammals. By the early Permian, they were the dominant terrestrial species. By the mid-Permian, another group of proto-mammals—the possibly warm-blooded, somewhat rodent-like therapsids—emerged and became the new dominant species. By the late Permian, they may have even developed fur. One group, known as *Theriodontia* (Latin for "beast tooth"), displayed a number of evolutionary innovations: A shift in the bones supporting the jaw allowed the jaw to open wider, and may have aided hearing as well. The skull and teeth became larger, the teeth more specialized, and the jaw more powerful. They seemed primed to take over. But it was not to be.

Everything changed at the Permian-Triassic boundary. Levels of CO_2 spiked, for reasons we'll discuss shortly. That led to massive warming. Plate tectonics by now had brought all the continents together into a single giant continent—Pangea—straddled across the equator. It was already difficult for maritime moisture to penetrate deep into the center of the continent. Rapid greenhouse warming made it even hotter and drier, according both to climate model simulations of the end of the Permian and analyses of the fossil river deposits from Pangean floodplains. The sudden drying would have led to the massive die-off of the tenuous, moisture-dependent forests that had arisen over the course of the Paleozoic. That meant less burial of organic matter on land, assisted perhaps by decreased carbon export to the deep oceans due to a collapsing marine food web. Atmospheric oxygen levels appear to have dropped precipitously as a result, reaching concentrations as low as fifteen percent at the P-T boundary.[3]

The drop in oxygen was a further contributor to the mass die-off. The combination of greenhouse warming and low oxygen would have led to widespread hypoxia—a state where organisms simply cannot

take in enough oxygen to support metabolism. That's where the dinosaurs come in. The proto-mammals that had come to dominance
during the Permian—the synapsids and therapsids—had thrived
off high oxygen levels. But as oxygen concentrations dropped, they
were now poorly suited to their environment. Enter the diapsids, a
wide-ranging group of tetrapod vertebrates that first emerged during
the Carboniferous around 300 million years ago. They include the
reptiles, birds, and now-extinct dinosaurs. What distinguished them
from the synapsids and therapsids was the evolutionary development
of two holes on each side of their skull. One subgroup of diapsids,
known as archosaurs—which includes crocodilians and the earliest
dinosaurs—exploited that innovation to develop a more efficient respiratory system that could make more effective use of the available
oxygen. That gave them a leg up on the competition when oxygen
levels plummeted at the P-T boundary. Dinosaurs, it turns out, were
direct beneficiaries of the P-T extinction event.[4]

Only a handful of proto-mammals survived. One group that did
was known as *Cynodontia* ("dog teeth"). They were our ancestors,
and the ancestors of all mammals. At first, they probably looked
somewhat like a huge, scaly rat, growing to as much as six feet in
length. Truly a Rodent of Unusual Size if ever there was one. But by
the end of the Triassic, they had shrunk to the size of modern-day
field mice, hiding behind rocks from their reptilian predators.

Carbon Catastrophe

Let's look at the CO_2 spike that triggered the cascade of events that
resulted in the greatest mass extinction event on record. There is no
evidence that the proto-mammals and early dinosaurs were building
coal-fired power plants or driving SUVs. What we do see is a series of
enormous volcanic eruptions originating from what are known as the
Siberian Traps, a hilly region in Siberia formed by repeated eruptions
of magma. This igneous province certainly is a good candidate for a
massive ancient release of carbon.

MIT geologist Sam Bowring and his team collected rocks from
a region in China that contains fossil evidence from the P-T extinction event. They used a dating method known as uranium-lead

geochronology, which takes advantage of the radioactive nature of uranium. Because the amount of uranium decreases over millions of years, the ratio of uranium to lead trapped in a crystal found in ancient rock can be used to estimate the age of the rock. Bowring and colleagues also collected rocks from the Siberian Traps and applied the same dating method to the layers of rock to estimate the beginning and end of the eruptions. Both the eruptions and the fossils dated to—you guessed it—around 250 million years ago. If science were a court of law, Siberian Trap volcanism would be convicted of causing the P-T mass extinction.[5]

The convict, it turns out, actually threw a one-two punch. Initially, the eruptions injected large quantities of volcanic ash into the atmosphere. That would have blocked out sunlight, reduced photosynthesis, and bombarded the surface of the planet with sulfuric acid rainfall. The planet would have resembled an apocalyptic hellscape for years following each eruption. The more persistent impact, however, was the accumulating atmospheric CO_2 as volcanoes continued to spew the gas into the atmosphere for tens of thousands of years.

We can estimate the rise in CO_2 levels that occurred a few different ways. Each involves the use of what are known as proxy data: natural biological, chemical, or physical recorders of past environmental change. It is important to look at what each of these lines of evidence suggests because to the extent they agree, we can draw more robust conclusions. But to the extent they disagree, we're faced with the inevitable "fuzziness" of drawing inferences about things that happened hundreds of millions of years ago.

The first source of proxy evidence is based on small pores, or stomata, in preserved ancient leaves. Plants open their stomata to let in CO_2, which they employ in photosynthesis, but there's a trade-off because that allows precious moisture to escape into the atmosphere. So when there is lots of carbon dioxide around, plants can make do with fewer stomata, allowing them to conserve moisture. Scientists can thus infer past changes in CO_2 by measuring this effect, though it's not without uncertainties.[6]

Another proxy method measures the relative abundance of the two main stable isotopes of carbon (carbon-12 and carbon-13) in carbonates found in ancient soils. Plants prefer to take up the lighter isotope (carbon-12) in photosynthesis, so the organic carbon that gets buried has less of the heavier isotope (carbon-13). Fossil fuels are buried organic carbon, so one of the ways we can establish that the rise in carbon dioxide we've seen over the past two centuries is due to fossil fuel burning is the fact that the atmospheric CO_2 is becoming increasingly depleted in carbon-13.[7]

Because atmospheric carbon dioxide is always significantly less depleted in carbon-13 than organic carbon, the soil near the surface, and most in contact with the atmosphere, is relatively un-depleted, whereas the soil below becomes increasingly depleted with depth. University of Utah geochemist Thure E. Cerling developed a method to estimate past changes in atmospheric carbon dioxide by measuring this effect. The method is not without its own assumptions and limitations, and indicates that other players, such as methane, might have been involved, too.[8]

A third and final proxy method also makes use of carbon isotope data but looks at the fossil remains of plants rather than ancient soil carbonates. Recent experiments demonstrate that the degree to which plants prefer to take up carbon-12 in photosynthesis depends on the concentration of CO_2. Higher carbon dioxide levels in the air lead to a greater depletion of carbon-13. So, by measuring the relative abundance of carbon-12 and carbon-13 in preserved ancient plant remains, one can once again estimate carbon dioxide levels.[9]

It would be nice if these methods all gave the same exact result. But science isn't always as clean as we might like it to be, and they don't. The strength of drawing upon multiple approaches, however, is that different methods make different assumptions, and though some of those assumptions might be questioned, the strengths and weaknesses of the different approaches are likely complementary in nature. By considering these lines of evidence collectively, we get a more robust picture.

What is that picture? When we take all of this paleo data collectively, we see a range of estimates of the rise in CO_2 levels at the peak of the P-T event. Carbon dioxide likely underwent a six-fold increase (2.5 doublings). But it might have been as little as a two-fold increase (one doubling) or as much as a twenty-four-fold increase (roughly 4.5 doublings). The lower end of the range corresponds to what we might see by the end of the century in a scenario of business-as-usual fossil fuel burning, whereas the upper end of the range is considerably greater than we're likely to see result from human activities.

What about the resulting warming? Here, too, the paleo data offer clues. First, we look at the relative abundance of the two stable isotopes of oxygen—oxygen-16 and oxygen-18—in calcium carbonate rocks and shells formed in seawater. The quantity of those isotopes varies with temperature, with warmer waters producing carbonates that are more depleted in the heavier oxygen-18. So, we can analyze the oxygen atom ratios in ancient carbonate rocks or calcite shells from ancient seabeds to estimate ocean temperature changes. Now, ocean salinity can also affect these ratios, but we can employ another method that measures carbon-oxygen bonds, also a function of temperature. Perfect measurements are hard to make, but these collective data paint a picture.

What we learn from all this is that prior to the P-T extinction event, tropical sea surface temperatures ranged from about 72°F to 77°F. During the peak extinction, tropical temperatures rose up to about 86°F. That's a warming of 9–14°F. As land tends to warm more than ocean, and the extratropics warm more than the tropics, the global average surface temperature might have warmed even more than that. During the past century, for example, global temperatures have warmed up about thirty-three percent more than tropical sea surface temperatures. If we assume a similar relationship today, that translates to a warming of 13–20°F.[10]

The warming was closely correlated with the rise in CO_2, making it likely that this was a greenhouse-driven warming event, similar to what we're experiencing today. It took place over 75,000 years—brief from a geological standpoint, but long in comparison to the current warming. That amounts to a warming of about 0.02°F per century. We're warming the planet now by about 2°F per century—a rate that's one hundred times greater.[11]

We now have some idea of both the increase in CO_2 and the amount and rate of warming that appears to have resulted from it. We can use this ancient episode to see how the warming effect of CO_2 applies to what we're seeing today.

Climate Sensitivity

There are several important lessons we can learn from this past episode. One of them involves the so-called sensitivity of Earth's climate to increasing greenhouse gas concentrations. That sensitivity determines just how rapidly we're exiting the bounds that define our fragile, habitable moment.

Climate sensitivity is defined in a very specific way: it's how much warming you get if you double the concentration of CO_2 in the atmosphere. Why doubling? Because the more CO_2 you add, the more you close the atmospheric "window" through which heat can escape to space. That means that it takes exponentially more CO_2 to create the same amount of warming. So, the warming that occurred between the last ice age, when CO_2 levels were about 180 ppm, to the 1990s, when they were about 360 ppm (a net increase of 180 ppm), is roughly the same as the warming that will happen if we double CO_2 from its current level of 420 ppm all the way to 840 ppm (a net increase of 420 ppm). This is how we arrive at the equilibrium climate sensitivity, or ECS, which is defined as the warming reached after the climate system fully responds to a doubling of CO_2 concentrations—for example, from preindustrial levels of about 280 ppm to 560 ppm, the level we'll reach mid-century under business-as-usual fossil fuel burning. That warming—note that because climate sensitivity is by convention always defined in Celsius, you'll have to forgive me for going back and forth between Celsius and Fahrenheit for this one quantity—was estimated as being between 1.5°C and 4.5°C (2.7°F and 8.1°F) back in the late 1970s in a National Academy of Sciences assessment led by the great MIT atmospheric scientist Jule Charney. That same range basically stands today. At the lower end of that uncertainty range, we would be looking at climate disruption that is qualitatively similar to what we've already seen. At the upper end, we'd be looking at catastrophic consequences: massive coastal inundation, withering summer heat, devastating droughts, and monumental floods. There is uncertainty—but how lucky do we feel? How risk-tolerant are we willing to be with our one and only planetary home?[12]

It's critical to recognize the key role played here by feedback mechanisms. In the absence of amplifying feedbacks, that is, if the only factor was the increase in the atmospheric greenhouse effect, the value of ECS would be around 1°C (2°F). We would only get 2°F of warming for a CO_2 doubling (we can call this the *no feedback* case). That wouldn't be so bad. Unfortunately, the reality is that key amplifying feedbacks kick in once we warm the planet. Melting of ice leads to more warming, our old friend the ice albedo feedback. Heating the oceans causes more evaporation of water vapor (a very potent greenhouse gas) into the atmosphere—that's the water vapor feedback. Both of these feedbacks add substantially to the warming.

Then there are clouds. High, wispy cirrus clouds provide an amplifying feedback because, like greenhouse gases, they block outgoing heat energy. Thick, low stratus clouds provide a stabilizing feedback because, like ice, they reflect outgoing sunlight back to space. And it's difficult to predict precisely how each type of cloud will change in a warmer climate. This remains the largest uncertainty among the key feedback mechanisms, and the primary source of uncertainty in estimating the precise ECS of the climate system.

There are other wild cards here, particularly so-called carbon cycle feedbacks. One is methane. Warming of permafrost and continental shelves can release previously trapped methane. As methane is a potent greenhouse gas, that can add to the warming, though in today's strongly oxidizing atmosphere, its lifetime is relatively short—decades, rather than centuries or millennia—and there's much less of it around than there is CO_2.

There are other carbon cycle feedbacks, though. Warming and drying can cause more extensive wildfires, like we've seen in Australia, western North America, and Eurasia in recent years, releasing carbon that was trapped in forests into the atmosphere. During the Australian Black Summer of early 2020, which I witnessed in person during a sabbatical in Sydney, more carbon was released from forests destroyed by bushfires than is produced from a year's worth of Australian fossil fuel burning.[13]

In any case, the collective evidence, from modern observations, paleo observations, and climate models, demonstrates that destabilizing feedbacks win out over stabilizing feedbacks. Though the ECS with no feedbacks at play is about 1.5°C (2.7°F), destabilizing feedbacks bring it closer to 3°C (5°F), and perhaps a bit higher. It's not a trivial difference—it's the difference between "we can easily adapt" and "we're in for a world of pain if we don't act."

So far, we've just considered the *fast* feedbacks. There are also slower feedbacks, like the stabilizing effect of increased weathering and CO_2 drawdown that plays out on geological timescales. There is also the response of ice sheets, the migration of forests, and other responses of the Earth system that take many centuries to play out. These slow feedbacks lead to a more general concept of Earth system sensitivity, or ESS, which describes how much warming you ultimately get in response to a doubling of CO_2 *after* these slow-response feedback mechanisms fully unfold. That's the quantity we're often measuring when we look at geological climate responses that occur over hundreds of thousands of years, rather than shorter-term responses due to volcanic eruptions or historical carbon emissions.

The P-T event was such a slow response, and we can learn something about ESS from it. Carbon dioxide, as we've seen, probably increased about six-fold during that period. The planet likely warmed around 16°F, which gives us an ESS of about 6.3°F. That is a bit more than conventional estimates of ECS, which, as we saw earlier, are closer to 5°F, suggesting that enhanced slow feedbacks may have added to the warming. But that ESS estimate is actually lower than some previous climate model–based estimates, e.g., about 7.7°F derived from the mid-Pliocene, a period around three million years ago when CO_2 levels last appear to have been as high as they are today. That time period provides an important possible analog for current warming, and we'll discuss it in depth in a later chapter.[14]

In any case, given the uncertainties reported earlier, the ESS calculated from the P-T extinction event could actually be as large as 20°F or as low as 3°F. Such a wide uncertainty range seems to render this event useless when it comes to figuring out climate sensitivity. Not so, however. It's simply one data point. We can bring numerous past events into the fold to create the fuller picture. Paleoclimatologist Dana Royer, while a post-doctoral researcher at Yale working with Bob Berner and Jeffrey Park, made use of the much broader information available, spanning from the Cambrian period 540 million years ago and ending with the present. When they looked at the past record of CO_2 changes from paleo data and took into account the relationship between chemical weathering and temperature on these longer timescales, they were able to narrow that window quite a bit. They found that the true value of ESS is likely within the range of 3–10°F.[15]

So, we've narrowed the ESS range down even further. We see how the more data we can use, the more certainty we can gain. Though there is uncertainty with every climate episode, having multiple episodes to work with averages away some of the uncertainty. But when it comes to climate sensitivity, we can do even better. Using lots of other examples of climate responses from the past that we'll encounter later in this book, like the cooling that occurred during the Last Glacial Maximum 21,000 years ago or the cooling response of the climate to sulfur and ash from major volcanic eruptions in recent centuries, climate scientists are constantly able to adjust and more accurately assess climate sensitivity. Because of that, ECS is currently estimated to likely be between 2.3°C and 4.5°C (4°F and 8°F) and almost certainly between 2°C and 6°C (3.6°F and 11°F).[16]

To quote the Genie from the live-action Disney movie *Aladdin*, "there's a lot of gray area" there. Warming of 3.6°F would be disruptive, but warming of 11°F would be catastrophic. We can get a better understanding of what could have led to the higher-end "worst-case scenarios" during events like the P-T extinction and, by implication, what might possibly happen today. Could there be processes currently not in play, at least not yet, that might have enhanced the sensitivity of the climate back then?

Let's look at methane, for example—a greenhouse gas with a much-discussed role in current warming. We've seen that it played a critical role in keeping the early, dimly lit Earth warm. Could a methane spike have contributed to the P-T warming? There is some evidence from the carbon isotope data that substantial amounts of methane buried in shallow seabeds might have been destabilized and released into the atmosphere. Methane is a very potent greenhouse gas, even more so than carbon dioxide, so you get more warming on top of the initial CO_2-induced warming. And more methane release. In other words, another destabilizing feedback. Another vicious cycle.

That destabilizing feedback is limited, however, by the fact that methane is readily removed from the atmosphere through oxidation.

Indeed, as we saw in the last chapter, it was the rise in oxygen more than two billion years ago that seems to have cooled the Earth into a snowball by scavenging the early methane greenhouse. That scavenging limits its time in the atmosphere today to decades, rather than, as with CO_2, centuries to millennia. During the P-T event, however, oxygen concentrations were lower, and methane might have stayed around modestly longer in the atmosphere, giving it greater long-term warming potential. That would make it more important during the P-T extinction, a time of low atmospheric oxygen.

The Four Horsemen of the Strangelove Ocean

In the oceans, ninety-six percent of all life appears to have died out during the P-T extinction. That's about as close to total extinction as you can come. The post-Permian ocean is what is sometimes called a Strangelove Ocean—a reference to the iconic Peter Sellers character in the 1964 Cold War dystopian film *Dr. Strangelove*, which ends (spoiler alert) with global thermonuclear war and the presumed end of human civilization. (Nearly sixty years later, the film suddenly feels ever-more prophetic in light of recent geopolitical developments, including Russia's threat to use tactical nuclear weapons in its war on Ukraine.)

By the time we finish reviewing the totality of what happened to the oceans during the P-T event—which involved at least four different fundamental insults to life (warming, acidification, anoxia, and hydrogen sulfide poisoning)—the question on your mind won't be, "Why did ninety-six percent of ocean life die off?" It will instead be, "How did four percent actually survive?"

The massive atmospheric carbon spike was itself already a double whammy: it didn't just cause global warming, but also its evil twin, ocean acidification. The rising atmospheric carbon dioxide seeped into the ocean, forming carbonic acid, which dissolved creatures with calcium carbonate shells and skeletons, destroying coral reefs and killing off crustaceans like crabs and lobsters and mollusks like clams and oysters. Their demise, in turn, upset the entire ocean food chain, including fish populations and other marine life.

The extinction of calcareous organisms in the P-T event fossil record certainly provides circumstantial evidence that a major ocean acidification event contributed to the mass ocean die-off. There's more than circumstantial evidence, though. Carbon and calcium isotope proxy data seem to show a sudden increase in dissolved ocean carbon. Though exact interpretations of the geochemical data are complicated and debated in the literature, I believe that the evidence is strong enough to indicate that that is what happened.

The most convincing evidence comes from what might seem like an unlikely source: boron. Boron combines with sodium and oxygen to make a white, powdery substance known as borax. It is found in large quantities in ancient, evaporated sea- and lakebeds—like the ones in California's Death Valley, where borax is still mined today for use in toothpaste, cosmetics, paints, and herbicides. It can also be extracted from ancient carbonates deposited on the seafloor, and the ratio of the two main stable isotopes (boron-11 versus boron-10) mirrors the pH of the ancient seas from which they came. One team of scientists found boron isotopes from ancient sediments, in the region that is now the United Arab Emirates, that date to the late Permian and early Triassic, when it was connected to an ocean neighboring ancient Pangea. They were able to reconstruct the change in ocean pH across the P-T boundary, finding that it remained stable during the early part of the P-T event due to low atmospheric CO_2 and high initial ocean alkalinity, allowing the ocean to buffer the relatively slow initial input of carbon. But it was unable to buffer the larger and more rapid injection that came later in the P-T event, suffering substantial, abrupt acidification—a drop of nearly one pH unit in just 10,000 years. It is primarily that rapid acidification event that drove the extinction of calcareous biota. And herein lies a cautionary tale. When it comes to our own precarious moment, we are once again subject to the twin threats of warming and ocean acidification. The combination of bleaching events from warming waters and ocean acidification threatens one of the natural wonders of our modern world, the Great Barrier Reef—as well as coral reefs around the world and other sea life upon which we rely for food.[17]

It is not only the amount of carbon we're adding to the atmosphere but the *rate* at which we're adding it that is problematic. The more rapid the increase, the worse the acidification of the ocean waters. Though the injection of carbon into the atmosphere during the Great Dying might indeed have been rapid by geological standards, it was nothing compared to what we're seeing today. The Great Dying saw a six-fold increase in carbon dioxide concentrations over 75,000 years. That amounts to a roughly 30,000-year doubling time. But we've increased carbon dioxide concentrations from preindustrial levels of 280 ppm in the atmosphere to current levels of roughly 420 ppm in about two centuries, amounting to less than a 300-year doubling time. In other words, we're adding carbon to the atmosphere a hundred times faster than the natural episode that caused the greatest extinction in planetary history. The P-T extinction event hints at some further worrying possibilities.[18]

The combined impact of warming and ocean acidification from rising CO_2 constituted a double whammy, but add in ocean hypoxia and you've got a triple whammy. We know that low oxygen levels in the atmosphere contributed to the P-T extinction of terrestrial organisms, but low oxygen levels in the atmosphere imply low oxygen levels in the ocean, too. That would have been compounded by the fact that warm ocean surface waters absorb and hold less oxygen (for the same reason that a warm bottle of soda wants to release its dissolved gases—hence the "decarbonation" when you open the top). An analysis of marine fossils at the P-T boundary shows that organisms with a low tolerance for hypoxia indeed had the highest extinction rates.[19]

Ocean surface warming also leads to greater stratification because lighter warm water wants to sit on top of denser cooler water. That leads to less turbulent upper ocean mixing, which means that oxygen and nutrients depleted by sea life are less likely to be replaced from rising, colder, more oxygen-rich waters. Research I've been involved in shows that such ocean changes are occurring today because of human-caused warming, and at a rate that is faster than models have predicted. These changes impact the entire marine food web. With fewer microbes and plankton, the small fish who feed upon them perish. Then the larger fish who feed on the little fish die off. And so on, all the way up the food chain to the apex predators. We end up with a collapse in marine productivity. Shades of the Strangelove Ocean.[20]

One study posits that a Strangelove Ocean, with less life in the upper ocean and less burial of organic carbon in the deep ocean, would have seen rising dissolved carbon in the upper ocean, and higher atmospheric CO_2 as a result, leading to further warming. Finally, reduced contrasts in temperatures between the poles and the equator, which are typical of relatively ice-free, greenhouse climates, lead to a more sluggish ocean conveyor and less delivery of oxygen to the deep ocean by the overturning ocean circulation. All of these factors together favored oxygen-depleted oceans at the P-T boundary. Climate model simulations have reproduced these conditions, lending more credibility to the theory.[21]

As if all of that wasn't enough, my former Penn State colleague Lee Kump and collaborators have argued for yet a fourth horseman in the apocalypse that was the P-T extinction event: a massive hydrogen sulfide "stink bomb." This prank came courtesy of Pele, the goddess of volcanoes (in Hawaiian mythology).[22]

Those of us who have been lucky enough to evade any actual stink bombs growing up will still recognize hydrogen sulfide gas as the smell of rotten eggs. Indeed, we can detect that smell in the most minute of amounts, in parts per *trillion*. In oxygen-depleted zones like the depths of the Black Sea, hydrogen sulfide is present in parts per *million*. Fed by massive input from rivers, the fresh waters in the upper part of the Black Sea are significantly less dense than those found in the saltier lower waters. That leads to highly stable stratification with oxygen remaining confined to the upper layers. Recent research has shown that the boundary between the oxygenated upper layer and the anoxic deeper layer has moved up in depth from about 450 feet to about 300 feet over the past half century, making the habitable region of the Black Sea forty percent smaller. The fish die-off is so severe that fishing fleets have now largely turned to sea snails and other mollusks. Warming-induced increases in stratification—a prospect we face for the world's oceans—are at least partially implicated for this trend.[23]

The disappearance of oxygen sets off a biogeochemical chain reaction. The organic debris produced by marine organisms is no longer consumed through aerobic processes, so bacteria that derive oxygen not from organic matter but from sulfur oxides tend to thrive. These bacteria produce hydrogen sulfide as they strip away oxygen. Hydrogen sulfide is poisonous to aerobic organisms, and the spread of hydrogen sulfide leads to larger and larger toxic zones. The nether regions of the Black Sea today are the poster child for this phenomenon. But could it have been more widespread in the past?

Kump and his coauthors make a compelling case that it was. As oxygen levels fell during the late Permian, they speculate that the increasingly anoxic ocean would have become filled with hydrogen sulfide. That would help explain the mass die-off of ocean life. It would also help explain the die-off of terrestrial animal and plant life, as the hydrogen sulfide came to the ocean surface, seeped out from the ocean, and filled the atmosphere with toxic levels of the gas.

An intriguing piece of evidence in support of this hypothesis comes from fossil spores from the end of the Permian. Deformities in these spores suggest increased exposure to ultraviolet radiation. Hydrogen sulfide destroys ozone, and an atmosphere filled with hydrogen sulfide—as hypothesized by Kump and colleagues—would lead to damage or even destruction of the ozone layer, another possible mechanism for the die-off and extinction of terrestrial organisms.[24]

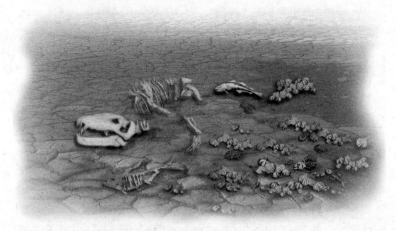

Figure 10. Artist rendering of the Great Dying extinction event.

Lessons Learned

The Great Dying is often pointed to as a potential analog for the consequences of current-day human-caused climate change. But it's an imperfect one. There was a cascade of environmental changes that were triggered by the massive eruptions of the Siberian Traps. Several appear relevant today, others less so. The message here is that there is cause for concern, and a strong reason to act. But it's certainly not a reason to give up hope for our species.

As we compare and contrast the P-T event with today, we see that certain threats portended by the great extinction loom over us today. Warming and ocean acidification are among them. Atmospheric anoxia and a global hydrogen sulfide "stink bomb"? Not so much.

The greenhouse-induced warming spike at the P-T boundary was undoubtedly an important contributor to the extinction event. But as we've seen, it played out about a hundred times more slowly than the current warming spike, giving life quite a bit more time to adapt to the effects of warming. Could this slower warming alone have triggered the massive extinction? My colleague Lee Kump, who has studied the P-T event as closely as any living scientist, is doubtful, concluding that it was "not large enough to cause mass extinctions by itself."[25]

That means that other aspects of this event were as, if not more, important contributors to the mass extinction as the warming itself. We know that the carbon cycle changes were critical to what unfolded, particularly the decreased organic carbon burial, which drove down atmospheric oxygen concentrations. A key development here was the dieback of forests due to hotter, drier conditions in P-T era Pangea. Forests were a relatively new ecological innovation and perhaps especially prone to the impacts of sudden warming and drying. The prevailing continental arrangement presented unique challenges. In the wake of the devastating Australian bushfires of 2020, which I witnessed up close, I remarked that "if you were going to pick the worst continent to live on as the climate changes, it would be Australia." It's a large continent centered in the warm, dry subtropics. Pangea, however, was even worse. It was much larger—occupying a whopping thirty percent of the total surface area of the entire planet (Australia, for comparison, occupies just one percent), with limited penetration of maritime moisture into its huge interior. And it was centered right on the hot equator.[26]

Today's forests are probably not as vulnerable to climate change. But that doesn't mean that they're immune to it, either. The Australian bushfires in the summer of 2020 burned nearly fifty million acres of forest, roughly the same area as the country of Syria, generating an approximate two percent increase in global CO_2 concentrations. Wildfires from the Amazon to the Arctic are releasing billions of tons of CO_2 a year. A study from 2020 in the leading journal *Nature* found that peak carbon uptake by tropical forests occurred during the 1990s and has declined ever since as a result of logging, farming, and the effects of climate change. At the current rate of deforestation, the world's largest rainforest—the Amazon, which has lost more than 600 million acres over the past four decades—could switch from a net absorber of carbon (a sink) to a net producer of carbon (a source) over the next decade. That's sooner than climate models have predicted.[27]

Though deforestation is a legitimate concern today for many reasons—not the least of which is forests' ability to store carbon—a climate-driven global collapse of forests and concomitant drop in atmospheric oxygen appears unlikely. Atmospheric oxygen levels are slowly dropping today, but that is tied directly to combustion from fossil fuel burning. It would take tens of thousands of years to drop from current concentrations (twenty-one percent) to P-T levels (fifteen percent)—not an immediate threat, but a reminder of the potential longer-term consequences of fossil fuel burning.[28]

What about *oceanic* oxygen depletion? We are certainly seeing decreased oxygen content in the world's oceans due, if for no other reason, to the fact that warmer waters hold less dissolved oxygen. Climate models underpredicted the observed decrease by about fifty percent. Some of my own research suggests that the models are failing to fully capture the increase in ocean stratification, which inhibits the mixing of oxygen into regions where it is depleted by respiring sea life. But we aren't witnessing anything close to what happened during the P-T extinction. Once again, leading expert Lee Kump suggests that the P-T event is a flawed analog for today because atmospheric oxygen isn't nearly as low and "today, there are not enough organics in the oceans to go anoxic."[29]

So, widespread decreases in atmospheric oxygen, ocean anoxia, and any global hydrogen sulfide "stink bomb" that might go with it, are less a threat today than during the P-T extinction event. They appear unlikely, by themselves, to threaten mass extinction now.

We're not done, though. What about ocean acidification? It, too, clearly contributed to the widespread ocean extinctions of the P-T event. We know that the pH of the ocean is decreasing at a rate of about 0.1 pH units a century. That might seem small, but it's not. Because pH is a logarithmic scale, that amounts to a thirty percent increase in acidity. During the height of the P-T event, pH might have decreased about 0.7 pH units in about 10,000 years. That's a rate of about 0.007 units a century, more than ten times slower than today, and yet it clearly had a detrimental impact on ocean life then.

We're already witnessing the damaging effects of ocean acidification today. Phytoplankton, mollusks, and crustaceans are encountering increasing difficulty forming calcareous shells and skeletons with decreasing ocean pH. Coral reefs around the world are under assault, too. The poster child is the Great Barrier Reef off the northeast coast of Australia where half the reef has now been lost. I saw the devastation myself, snorkeling with my family near Cairns, Australia, in late 2019. The reefs of the closer-to-home Caribbean, which I've seen close up, too, are also being decimated by ocean acidification.[30]

Given business-as-usual carbon emissions, the pH of the ocean is projected to drop to the point where corals can no longer form aragonite (a form of calcium carbonate) skeletons within a matter of decades. Cold water reefs in regions such as Scotland, Norway, Patagonia, and the Antarctic are similarly threatened. Reefs are an important source of ocean biodiversity and provide a protective habitat for many fish species, and their loss, along with the loss of plankton and shellfish, threatens the global ocean food chain.[31]

It's not just prawns and porpoises but people who are imperiled. The shellfish industry in the United States, from New England to the Pacific Northwest to the Gulf of Mexico, has been negatively

impacted by acidification. Harmful algae species like those responsible for toxic red tide blooms prefer lower pH environments. Red
tides are a health threat to us humans, as they can contaminate the
entire ocean food chain, including the seafood we consume. A collapsing ocean food chain threatens human civilization, given that
roughly one fifth of the global population relies primarily upon seafood as their major source of protein. When it comes to the threat
of ocean acidification, the P-T extinction event indeed constitutes a
cautionary tale.

And what about ozone depletion? We saw that a hydrogen sulfide
"stink bomb" might have damaged or even destroyed the ozone layer
during the P-T event. Such a scenario is highly unlikely today. But
we have depleted stratospheric ozone through other means, namely
the industrial production of ozone-destroying chemicals known as
chlorofluorocarbons (CFCs) and Freons. By the 1970s atmospheric
scientists were measuring substantial loss of stratospheric ozone in
the Southern Hemisphere, where the very strong band of stratospheric
winds known as the circumpolar vortex concentrates the damaging
chemicals and accelerates ozone depletion. Ozone depletion has also
been observed, albeit to a lesser extent, in the Northern Hemisphere.
Among the adverse consequences are human health ailments, such
as skin cancer and cataracts, and damage to terrestrial and aquatic
ecosystems and food chains. In 1987, the international agreement
known as the Montreal Protocol—which was signed by then U.S.
president Ronald Reagan—banned the environment-damaging
chemicals. Though we're not thoroughly out of the woods yet, some
observers have defensibly classified the ozone hole as an example of
"when the world actually solved an environmental crisis."[32]

Last but certainly not least, what does the P-T extinction event
have to say about prospects for runaway methane-driven warming?
The event is sometimes cited as evidence of just that. Some protag-
onists, like scientist-turned-doomsayer Guy McPherson, insist that
such a scenario is underway now—that large reservoirs of previously frozen methane have been mobilized by the warming of the
Arctic, triggering a runaway, amplifying feedback process by which
more methane begats more warming and more melting, and yet more
methane until all of it—roughly fifty billion tons—is released into the

atmosphere. Such a scenario, they argue, could double the amount of warming we face. McPherson goes even further, insisting not only that this is already happening, but that it *will* lead to the imminent extinction of all life on Earth. The evidence does not remotely support such contentions. Methane increases are being observed, but recent studies show that the rise in methane is tied to natural gas extraction, livestock, and farming. In other words, it's not part of some runaway warming loop. It can be stemmed if we take action, just the opposite of what the doomist hot takes imply.[33]

What does the P-T event tell us about methane feedbacks specifically? One problem is that methane would likely have persisted longer in the atmosphere during the P-T boundary than it does today, owing to the rather low concentrations of oxygen. That's another "non-analog" complication in drawing direct inferences with regard to modern warming. The pre-extinction Late Permian seems to have been a slightly warmer climate than today with slightly higher CO_2 concentrations. There were no ice sheets around and relatively little if any ice on the planet at all. Yet even with this warmer baseline climate state, we didn't see runaway warming in response to the P-T carbon spike. Tropical sea surface temperatures may have climbed to 86°F, which is about 9°F warmer than today. Global average surface temperatures were probably closer to 77°F, and the net global warming was about 16°F. That, as we have seen, is in line with expectations given standard estimates of Earth System Sensitivity. Admittedly, there is a very wide uncertainty range in those estimates, but neither is there any compelling evidence for missing mechanisms, runaway methane-driven warming, or any other such exotic inferences. The warming was basically what would be expected. And that's bad enough.

The P-T event is the most pronounced extinction event in the geological record. But might there be other mass extinction events from which we can learn some valuable lessons about the predicament we face today? Indeed there are, and none is more famous than the one that killed off the dinosaurs—the so-called K-Pg extinction event that marked the end of the Cretaceous period. It's the subject of our next chapter.

4

Mighty Brontosaurus

Don't You Have a Lesson for Us?

Hey, mighty brontosaurus.
Don't you have a lesson for us?
You thought your rule would always last.
There were no lessons in your past.
You were built three stories high.
They say you would not hurt a fly.
If we explode the atom bomb,
Would they say that we were dumb?

—GORDON SUMNER (AKA STING),
"Walking in Your Footsteps"

Do the dinosaurs, victims of a famous sixty-six-million-year-old mass extinction event, have a message for us? That rhetorical question was posed by the rock band The Police in their 1983 song "Walking in Your Footsteps," which came out during my junior year in high school. What I and most listeners weren't aware of then was that this evocative track off the album *Synchronicity* was actually a parable about the Cold War, nuclear holocaust, and—though The Police themselves may not have intended it as such—catastrophic climate change.

AS WE WILL SEE, there are remarkable parallels indeed between the dinosaur-killing asteroid impact and the Nuclear Winter worries of the 1980s. There is a direct scientific connection via a shared scenario of global climate change (cooling, rather than warming, in this case). But there are lessons here about the nature of scientific understanding and how it evolves, the self-correcting nature of science, the tactics used by vested interests to undermine public faith in science disadvantageous to their business model, and how all of this impacts the public discourse over planetary environmental threats—be they the prospect of Nuclear Winter we faced in the 1980s or, more germane to the subject of this book, the climate crisis we face today.

Of Bombs and Bolides

In fall 1984, just a year after "Walking in Your Footsteps" was released, I arrived at the University of California at Berkeley to pursue an undergraduate degree in physics. I had been lured there, in part, by the prospect of learning from the school's scientific luminaries, which included a dozen Nobel laureates. A few years earlier, in 1980, one of their Nobel Prize–winning physicists, Luis Alvarez, and his geologist son Walter had made an Earth-shattering discovery. Literally. They had analyzed layers of deep-sea sediment around the world that all dated to the transition between the Cretaceous period and the Paleogene period, known as the K-Pg boundary (it was once referred to as the K-T boundary). The K-Pg deposits showed evidence of a massive impact event, containing levels of iridium—a hard, brittle, silvery-white metal that is extremely rare in Earth's crust, but common in extraplanetary asteroids.[1]

The K-Pg boundary was no run-of-the-mill geological boundary. It coincided with the mass extinction event that ended the dinosaurs' own fragile moment. Having been ushered in by the P-T extinction event 250 million years ago, they were exterminated by the K-Pg extinction event. Live by the mass extinction, die by the mass extinction. There are always winners and losers. And the winners can't win 'em all. This time the dinosaurs were the losers, and the beneficiaries were us, the mammals.

Sediments below the K-Pg layer contain fossils of iconic therapods like *Tyrannosaurus rex*. Sediments above that layer? Not only are dinosaur fossils completely absent, but there are no fossil remains of any large terrestrial species (specifically those estimated to have weighed more than about fifty pounds). Unlike smaller reptiles and early mammals, large creatures were unable to shield themselves from the ensuing environmental Armageddon by hiding in burrows. Roughly three quarters of all animal species perished. Some mass extinctions—as we saw with the P-T event of the last chapter—are complicated, with several factors or combinations of factors possibly at play. This one is simple. An asteroid killed off the dinosaurs and every amphibian, reptile, and mammal larger than a bulldog. Understanding the precise details of what unfolded, and how that relates to our current moment, will nonetheless require a bit more digging.[2]

If this were a murder case, the critical piece of evidence missing at the time of the bombshell Alvarez discovery in 1980 was the weapon. It had been found in the 1970s, when geophysicists searching the Yucatán Peninsula for oil (it always seems to come back to fossil fuels) had uncovered evidence of a massive, roughly 100-mile-wide crater, known as Chicxulub.

At the time, these geophysicists didn't recognize the crater as the aftermath of the impact by a bolide—in this case, an asteroid. It wasn't an oversight—craters can come from other sources, like volcanic eruptions. Only in 1990, a full decade after the Alvarez discovery, did other scientists nail that down. Geoscientist Alan Hildebrand and colleagues uncovered geological and geochemical evidence from Chicxulub—including exotic minerals, such as shocked quartz (a type of quartz that is only formed under extremely high pressure), and geological formations suggestive of a huge shock wave—that the crater was indeed formed by a very large, roughly six-mile-wide asteroid. And the sediments dated to the K-Pg boundary. Weapon found. We have a conviction.[3]

It is hard to imagine what this event might have looked and felt like, for there are simply no modern analogs for such a cataclysmic event. The asteroid struck Earth with the almost inconceivable force of somewhere between a hundred million and ten billion Hiroshima

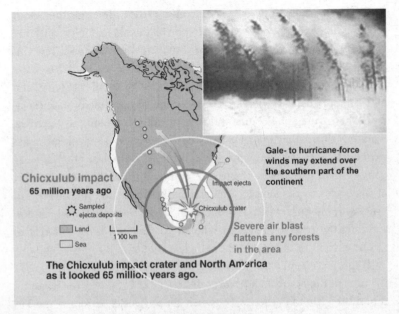

Chicxulub impact
65 million years ago

⬡ Sampled
   ejecta deposits

⬜ Land   └─────┘
                 1 000 km
⬜ Sea

Impact ejecta

Chicxulub crater

Gale- to hurricane-force
winds may extend over
the southern part of the
continent

Severe air blast
flattens any forests
in the area

The Chicxulub impact crater and North America
as it looked 65 million years ago.

Figure 11. Depiction of the Chicxulub impact event.

bombs. Evidence of the aftermath has been collected by scientists over several decades, and it does provide some remarkable glimpses of what unfolded. Scientists, for example, have detected geological structures formed by seismic waves that emanated outward from the impact. The concentric ripples traveled through soil and rock as if it were warm maple syrup. Massive amounts of dust, soot, and particulate matter known as aerosols were blasted into the atmosphere. Seismic imaging reveals evidence of mile-high tsunamis that assaulted the Gulf Coast of North America. Supercomputers have recently been used to simulate the physics of the collision, suggesting that the asteroid probably struck at an angle of about sixty degrees, rather than a right angle. That would have maximized the ejection of debris into the atmosphere, a worst-case scenario for the dinosaurs.[4]

Our best window into the event yet, however, was provided recently, in the form of the remains uncovered by a team of paleontologists at the Tanis site in North Dakota, about 2000 miles away from the asteroid impact. We typically talk about what fossil remains can be found before the event or after the event. But in this very rare case, fossils—which were exquisitely preserved in layers of mud and tree resin—appear to have actually been directly tied to the event itself. The jumbled remains of animals and plants, both terrestrial and aquatic, seem to have been caught up in wave-like torrents of water excited by the seismic tremors created by the impact event. The leg of a *Thescelosaurus* (meaning "wonderous lizard")—a small, roughly ten-foot-long, herbivorous, bipedal dinosaur with a sturdy build, small forearms, long pointed snout, and large bony eyebrows—was found, still covered with skin. The skin of a horned triceratops was recovered, scales intact. There's an embryo of a flying pterosaur, still inside its egg. A turtle fossil was found, impaled by a wooden stake. So, too, were some of our small mammal ancestors, along with burrows they had dug. Most remarkable of all, there appears to be a fragment from the asteroid itself. This is, of course, just one site where scientists were fortunate enough to uncover what happened at the moment of the K-Pg event. Imagine scenes like this occurring at the same precise moment across North America, South America, Eurasia, Africa, and Australia. Now we have some sense of what this event might have been like.[5]

The 1980 discovery that an asteroid had killed the dinosaurs was remarkably timed. It was the Reagan era, the height of tensions between the Americans and the Soviets. There was an escalating nuclear arms race between the two nations, and mankind was threatened with the prospect of a devastating global thermonuclear war.

As suggested by the lyrics that began this chapter, there was an eerie connection between these two events: the asteroid that doomed the dinosaurs and the existential threat posed by nuclear war. One might call it *synchronicity*—the Jungian concept that two events disconnected in space and time can be meaningfully related while lacking any direct causal connection.

Nuclear Winter

The massive dust and debris ejected into the atmosphere by the K-Pg asteroid impact blanketed the planet. Massive terrestrial wildfires were ignited by the burning debris that was ejected, generating

Figure 12. Sir David Attenborough examines the preserved scales of a horned triceratops recovered during the Tanis excavation.

substantial amounts of smoke and soot. The dust, debris, and soot remained suspended in the atmosphere for years, reflecting a sizeable portion of incoming sunlight back to space, and sending the planet into a sort-of perpetual winter while also severely limiting photo-synthesis—a double whammy for life on Earth. Larger land animals who could not burrow into the Earth, to limit their exposure to either any initial fires or, more importantly, the subsequent chill, selectively perished. The dinosaurs were among the casualties.[6]

So, ironically, it wasn't global warming but global cooling that de-livered their fate. Herein lies a key observation: what creates vulnera-bility to climate change isn't so much the precise nature of that change (for example, cooling versus warming), but what conditions species are accustomed to. Plants and animals can adapt to slow changes. Very rapid changes, not so much. Today, the rate at which climate zones are shifting poleward with warming is exceeding the rate at which many plants and animals can move, threatening (along with deforestation, habitat destruction, pollution, ocean acidification, and other activities) a human-caused *sixth extinction* event. If climate zones were shifting equatorward instead, due to equally rapid global cooling, animal and plant species would be similarly threatened. It's not just the magnitude of change but the rate of change that matters here.[7]

What is the relevance of all of this to global thermonuclear war? Obviously, a massive bolide strike and a massive array of nuclear detonations would both cause great physical destruction—they clearly have at least that much in common. Popular depictions of thermonuclear holocaust in the early 1980s portrayed the widespread physical devastation that would be rendered—entire cities destroyed, large populations of human beings in urban centers vaporized.

In summer 1983, just after the completion of my junior year in high school, my friends and I went to see the film *WarGames* in a crowded, hot movie theater in Hyannis, Massachusetts, during a week-long bike trip to Cape Cod. The protagonist, played by a young Matthew Broderick, is a nerdy yet brash teenage science prodigy who accidentally gets into the NORAD computer, Joshua, that controls the U.S. nuclear arsenal. He ends up triggering a nuclear escalation that appears to be headed toward mutually assured destruction. The computer assesses the impact of prospective strikes against Soviet population centers and the likely Soviet counterstrikes. Broderick's character saves the day by getting Joshua to play tic-tac-toe against itself and "learn" that there can be no winner, in either tic-tac-toe or a full-scale thermonuclear war. Joshua memorably exclaims, to the relief of those gathered, that "the only winning move is not to play."[8]

Just before the Thanksgiving holiday that year, *The Day After* (not to be confused with the much later, 2004 climate change–themed film *The Day After Tomorrow*) aired as a special on the ABC network. With a hundred million viewers, it was the most-watched TV film of all time. The film depicted a full-scale nuclear exchange between the United States and the Soviet Union, focused on residents of Lawrence, Kansas, and the collapse of civilizational infrastructure that occurs both in the lead-up to war and then in its aftermath. I remember my high school offering special counseling sessions for distraught students the day after the film. (This was in the progressive college town of Amherst, Massachusetts; my wife, who grew up in the more conservative community of Darien, Connecticut, remembers nothing of the sort.)

Both of these popular portrayals focused on the acute physical devastation and mortality that would be caused by an all-out nuclear war, adequate grounds on its own to end the Cold War escalation. Ronald Reagan himself wrote in his diary that he was inspired by *The Day After* to engage in more concerted efforts to achieve an arms agreement with Russia. But the impacts were for the most part depicted as fleeting. The implication was that if one managed to avoid the direct strikes and exposure to radiation, one could survive a nuclear holocaust. Both films were in production by 1982. By the time they appeared, in mid- to late 1983, a yet more dire picture had emerged.[9]

That's when Carl Sagan comes into the picture. If you're looking for more "meaningful coincidences," the name Carl—be it Carl *Jung* or Carl *Sagan*—is, according to numerologists, associated with "intellectuals, deep thinkers, philosophers and scholars," whose "souls are visionary and their way of thinking unique."[10]

Sagan and his collaborators had been researching the potential climate impacts of a global thermonuclear war. They found that the massive dust, wildfire smoke, and aerosols that would be ejected into the atmosphere in response to a series of thermonuclear detonations in the United States and Eurasia could trigger a nuclear winter. In December 1983, they published their findings in the leading journal *Science*. The publication has come to be known as TTAPS, named after the authors (Turco, Toon, Ackerman, Pollack, and Sagan) of the article. The team estimated that several thousand megatons of nuclear detonations would lead, within a matter of weeks, to an eighty percent reduction in sunlight reaching the surface of the Earth. They calculated that this would cause a cooling of 27–45°F over mid-latitude land regions, yielding subfreezing temperatures for months, even in summer. The combination of extreme cold, nuclear fallout, radiation exposure, and ozone destruction, they argued, could "pose a serious threat to human survivors and to other species." In other words, even those who survived the immediate physical devastation of a global thermonuclear war might suffer the same fate as the dinosaurs, perishing from the sustained deep freeze.[11]

Scientists had speculated about nuclear winter as early as the late 1970s. Paul Ehrlich; his wife, Anne Ehrlich; and their colleague John Holdren (who went on to become President Barack Obama's science adviser in 2009) discussed the possibility that the dust and smoke from a nuclear war could have a cooling effect similar to an explosive volcanic eruption, like the 1815 Mount Tambora eruption, back in 1977. In 1982, a year before the publication of TTAPS, Paul Crutzen—who would go on to share the 1995 Nobel Prize in Chemistry with Sherry Rowland and Mario Molina for working out the chemistry behind human-caused ozone depletion back in the early 1970s—coauthored an article ("The Atmosphere After a Nuclear War: Twilight at Noon") estimating that the smoke and aerosols from nuclear war–induced fires could reduce sunlight by as much as a factor of a hundred, dramatically decreasing photosynthesis with adverse consequences for agriculture. They concluded that nitrates produced by the massive fires would worsen both ozone depletion and acid rain. They drew a direct analogy with the K-Pg mass extinction event.[12]

Sagan and his TTAPS colleagues, however, were the first to quantitatively estimate the cooling impact of nuclear winter. Sagan perceived it such an imminent and existential threat to humanity that he chose to do something somewhat unusual (and often frowned upon) for a scientist: going public with his findings before the work was peer-reviewed and published. Launching his efforts in October 1983, several months before publication, Sagan set out to warn the public of the dire threat posed by the escalating arms race and nuclear winter. The publicity campaign included a high-profile press conference. It also included an October 30 commentary that served as the cover story of the widely read *Parade* Sunday newspaper insert, which had a circulation in excess of thirty million. The front cover featured an image of a devastated Earth half covered in shadow and sprinkled with icy white snow, posing the rhetorical question: "Would nuclear war be the end of the world?"

Sagan's actions rubbed some of his fellow scientists the wrong way. They were critical of some of the assumptions made in the TTAPS article and felt that the findings should have been litigated through the normal scientific process, rather than aired in public prior to publication. What started out as an internal scientific disagreement would soon spill out into the public arena. It would also play into the hands of Cold War hawks looking to discredit Sagan's call for nuclear de-escalation—and, ultimately, it would fan the flames of climate change denialism, as we'll learn later.

As we have seen, Sagan was assisted in his publicity efforts—if somewhat serendipitously—by The Police with "Walking in Your Footsteps," released in mid-1983, just months before Sagan's nuclear winter publicity campaign. The song was written by Sting one December 1982 morning on the Caribbean island of Montserrat. That makes the work eerily prescient because it preceded any widespread public awareness of the nuclear winter threat. Though the song was written about nuclear war, the analogy drawn between the effects of a global nuclear conflict and the demise of the dinosaurs is far more fitting than Sting presumably realized when he wrote the song.[13]

Scientist as Public Figure

Two books alerting me to the growing threat to our environment sat side by side on our family bookshelf when I was growing up in the early 1970s. One of them was Paul Ehrlich's *The Population Bomb*—a prescient, early warning of our collision course with environmental sustainability. The other was Rachel Carson's *Silent Spring*, which exposed the devastating environmental impact of the insecticide DDT. Both books were denounced by critics of the environmental movement. Julian Simon of the ultraconservative Cato Institute called Ehrlich an alarmist purveyor of doom and gloom promoting a "juggernaut of environmentalist hysteria." Decades later, Ehrlich's prognostications have nonetheless proven prescient. A group of more than 1500 of the world's leading scientists, including half of the living Nobel Prize winners, has confirmed that "human beings and the natural world are on a collision course" and that human activity is inflicting "harsh and often irreversible damage on the environment and on critical resources." The national academies of the world have weighed in with similar sentiments.[14]

I found myself subject to similar efforts to discredit me and my research following publication of the hockey stick curve with colleagues

Raymond Bradley and Malcolm Hughes in the late 1990s. The now-famous curve demonstrated the dramatic and unprecedented nature of modern human-caused warming. It was a threat to powerful vested interests, including the fossil fuel industry and its ideological allies in politics and the media. Paul Ehrlich was among the first colleagues to come to our defense.[15]

When it comes to scientists under fire, Carl Sagan is a canonical example. As we have already seen, he was a leading planetary scientist who made fundamental contributions to climate science, including his seminal work on the Faint Young Sun paradox. But he was far more than a scientist. He was the greatest science communicator of our time. He had a stage presence that was unmatched among scientific figures and a unique ability to engage the public on matters of science. Indeed, it was Carl Sagan and his epic thirteen-part *Cosmos* series, which premiered at the start of my freshman year in high school, that inspired me to pursue a career in science. To my great regret, I never got a chance to meet Sagan. He passed away in 1996 as I was completing my Ph.D. But I've had the pleasure of getting to know him through his writings, and to make acquaintance with some who knew him well, including his daughter Sasha, a talented writer who is continuing her father's legacy of communicating the wonders of science and reason to the world.[16]

Sagan was so charismatic and compelling a figure that he became the voice of science in America in the late 1970s and early 1980s. On Johnny Carson's *The Tonight Show* he regularly regaled late-night television audiences with his astute observations, insights, and often-amusing anecdotes. One early appearance in 1978 was especially memorable. Sagan complained to Johnny Carson and a stunned-silent audience about the chauvinistic and racially myopic depiction of the interstellar beings in the recent blockbuster film *Star Wars*: "They're all White . . . Everybody in charge of the galaxy seemed to look like us. I thought there was a large amount of human chauvinism in it." No medal is bestowed upon Chewbacca, despite his heroics, which Sagan called "anti-Wookiee discrimination." That last part was tongue-in-cheek and yet deadly on target at the same time. This parochialism might seem obvious to us now when pointed out, given our considerably more multicultural outlook, but Sagan was characteristically ahead of his time in providing us a more expansive view of our place in the world and the universe.

Sagan had become increasingly political in the 1980s as he recognized the mounting threat of the nuclear arms race. He used his public prominence, media savvy, and unrivaled communication skills to raise awareness about the growing threat. As Matthew Francis, writing for *Smithsonian Magazine*, put it: "Sagan, like many at the time, believed nuclear war was the single greatest threat facing humanity. Others—including policymakers in the Reagan administration—believed a nuclear war was winnable, or at least survivable. Making the danger of nuclear winter real to them, Sagan believed, would take more than science." Sagan even, as we've seen, took the unconventional—for a scientist—step of announcing his research on nuclear winter to the public even before the underlying science had been peer-reviewed and published.[17]

There are certainly times when such actions have proven unconstructive. The cold fusion episode is one cautionary tale. In March 1989, electrochemists Stanley Pons and Martin Fleischmann held a press conference to make the bombshell announcement that they had produced fusion at room temperature using a simple tabletop electrolysis apparatus. If true, it would have not only revolutionized our understanding of physics, but it would have seemingly provided a nearly endless supply of extremely cheap energy. But it was not to be. Within weeks of their announcement, numerous other groups had attempted and failed to replicate their results. The paper they had submitted to *Nature* was never published. Thirty years later, cold fusion is regarded as the definitive example of pathological science.[18]

But do extraordinary—or indeed, *existential*—threats not call for extraordinary measures? Consider, for example, Robert Jastrow's successor at NASA Goddard Institute for Space Studies (GISS), Dr. James Hansen, sometimes considered the "father of global warming" for his seminal work in the 1980s alerting the public to the threat of human-caused climate change. Raised a conservative, mild-mannered Midwesterner, Hansen in the latter part of his career turned toward activism and civil disobedience, participating in protests of mountaintop removal coal mines and oil pipelines. More recently, another NASA climate scientist chained himself to a JPMorgan Chase building in Los Angeles to protest what he sees as a lack of action on climate. Sagan's actions, by comparison, actually seem quite tame.[19]

Climate Wars—The Cold Variety

It is almost certain that Sagan's outspokenness, his prominence as a public figure, and his celebrity status rankled some of his scientific peers. It seems plausible—if not outright certain—that it got him blackballed from entry into the U.S. National Academy of Sciences, despite having made fundamental contributions to science that exceed those of most Academy members (myself included). That same resentment within the scientific community, furthermore, led to some particularly vitriolic challenges to his scientific work that were leveraged as a cudgel against him by his more politically and ideologically motivated critics.[20]

That brings us back to Sagan's (TTAPS) nuclear winter work. Sagan and colleagues had employed a somewhat primitive type of climate model called a radiative-convective model. It is similar to the elementary climate model we encountered in Chapter 2 in that it averages over the planet, treating Earth's surface as a single mathematical point and solving for the average surface temperature of the planet based on a balance between incoming and outgoing heating. However, it does provide more realism, accounting for the vertical dimension in the atmosphere. The same sort of model, in fact, was used by James Hansen in 1981 to study future global warming scenarios, demonstrating that historical global temperature changes could be explained by a combination of natural factors—such as volcanic eruptions—and the human factor of increasing carbon pollution. Hansen predicted that continued fossil fuel burning could lead to "potential effects on climate in the 21st century" that include "the creation of drought-prone regions in North America and central Asia as part of a shifting of climatic zones, erosion of the West Antarctic Ice Sheet with a consequent worldwide rise in sea level, and opening of the fabled Northwest Passage." Each of these things has since come to pass. Around the same time, fossil fuel giant ExxonMobil's own scientists had in fact secretly predicted the very same thing.[21]

Human-caused warming would increasingly come into focus in the late 1980s and early 1990s as the Cold War subsided and the threat of human-caused climate change mounted. But for the time being, the focus was on nuclear winter.

The TTAPS study employed this sort of model to focus on the catastrophic sudden worldwide *cooling* that would result from a global nuclear war. The vertical nature of the model allowed Sagan and colleagues to account for the effects of reflective dust and aerosols in the atmosphere, which are important features in the nuclear winter scenario. But the still-crude nature of the model earned Sagan and colleagues some criticism from fellow scientists. And not just by contrarians and curmudgeons.

Enter Stephen Schneider. Schneider was a Stanford colleague of Paul Ehrlich. He was the most articulate voice in the climate change debate, devoted to outreach efforts aimed at informing the public discourse over human-caused climate change, its impacts and solutions. Schneider—a contemporary of Sagan—was, if you like, sort of the "Carl Sagan" of climate change. He was also a greatly respected climate scientist. A member of the U.S. National Academy of Sciences, he made key early contributions to the science of climate modeling and performed some of the key early climate change experiments. Later in his career, he spearheaded efforts toward interdisciplinary climate science, including so-called *integrated assessment*—the science of coupling projections of climate change with models of potential climate change impacts, to inform real-world decision-making.

Schneider was also a mentor who provided me with critical advice when I was first under attack by climate change critics in the early 2000s. He tragically passed away from a heart attack in July 2010, just weeks before I was to see him at an event at Google. Among the personal recognitions I'm proudest of are the awards I have received (the Stephen H. Schneider Prize for Outstanding Climate Science Communication of the Commonwealth Club in California, and the Stephen Schneider Lecture of the American Geophysical Union) that honor Schneider's dual legacy as a scientist and science communicator.

Schneider, like Paul Ehrlich and like Rachel Carson before him, was the subject of attacks by contrarians seeking to discredit him and his scientific work. Back in the early 1970s, it was still unclear as to whether the warming effect of human-generated greenhouse gases or the cooling effect of industrial sulfur pollution from coal-fired power plants (which form aerosols that, like volcanic sulfate aerosols, have a cooling impact on the planet by reflecting sunlight back to space) would win out. Schneider, in an article he coauthored with NASA scientist S. Ichtiaque Rasool, speculated that the sulfates and cooling could possibly win out in the absence of environmental regulations.[22]

But the regulations came. Sulfur pollution was also behind the growing acid rain problem in the eastern United States, and Congress passed legislation (the Clean Air Act) requiring power plants to scrub the sulfur dioxide from smoke stacks before it escaped into the atmosphere. Those policies worked. They alleviated the acid rain problem, and they also unmasked the greenhouse warming that had been "hidden" by industrial sulfur pollution. To this day, climate contrarians continue to make the misleading claim that Schneider predicted cooling in the 1970s (he didn't—he simply alluded to the possibility). Based on this false premise, they pose the rhetorical question of why we should believe climate scientists about warming today? It's both a cynical smear and a dangerous gambit.[23]

Sagan and Schneider were both personal heroes and role models of mine—superb scientists who were outstanding communicators and who fought back against attacks and smears. So, I find it tremendously disappointing that the two of them didn't see eye to eye. In fact, they argued somewhat bitterly over nuclear winter. And they did so in a way that inadvertently played into the agenda of those looking to undermine public faith in climate science.

In 1986, Schneider coauthored an article with Starley Thompson of the National Center for Atmospheric Research in *Foreign Affairs*, widely read in policy circles. They took issue with the TTAPS findings based in part on their own research published just four months later in the competing journal *Nature*. The two argued that the scenarios explored in TTAPS of *thousands* of megaton nuclear exchanges were extreme (the most powerful nuclear weapon ever detonated was just *fifty* megatons) and likely led to an overestimate of the potential for massive fires, smoke, and dust. They used a slightly more sophisticated model that allowed them to account for other important variables, like atmospheric wind patterns and rainfall, that can impact the distribution of particulates in the atmosphere. In doing so, they arrived at a roughly twenty-five percent lower estimate of the overall cooling than TTAPS.[24]

A few years later, Sagan and the TTAPS team would essentially agree with the more modest Thompson and Schneider estimates. So, two different teams of experts converged in their estimates of a potentially catastrophic planetary cooling in response to a global thermonuclear conflict. Case closed, right? Wrong. The author of a *New York Times* article on the new "TTAPS2" paper reached out to Stephen Schneider for comment. His reply: "I would call it nuclear fall, not winter," adding, "but in any case, the TTAPS numbers have now more or less converged with ours, so I don't have a major problem with them anymore." In context, this was a reasonable, even conciliatory statement on Schneider's part, acknowledging that the basic premise of TTAPS was correct, even if the specific numbers could be quibbled with. The natural science communicator that Schneider was, he looked for a simple analogy that the public could grasp. The cooling effect was real but somewhat less than the original TTAPS "nuclear winter." "Nuclear fall" seemed to convey that notion.[25]

The savvy public figure that Schneider was, on the other hand, he might have anticipated how his comments could be played by those with an agenda. Did he mean to disparage Sagan's "nuclear winter" framing? Was there perhaps a bit of rivalry behind this? I don't think we'll ever know, and I'm not sure it matters. As Alan Robock, a climate modeler and nuclear winter expert, has noted, "they didn't mean for people to think that it would be all raking leaves and football games, but many members of the public, and some pro-nuclear advocates, preferred to take it that way," adding "the fight over the details of the modelling caused a rift between Sagan and Schneider that never healed." It also provided a huge opening for Cold War hawks looking to discredit what they saw as the real threat—Sagan and his open advocacy for nuclear disarmament.[26]

It's Personal

Sagan had become the public face of nuclear winter. It was Sagan who was invited to debate the topic before Congress in 1984. It was Sagan who was later invited by Pope John Paul II to discuss the topic. And in 1988, it was Sagan who was mentioned by name by Soviet premier Mikhail Gorbachev as his inspiration for ending nuclear proliferation during his meeting with Ronald Reagan. It was *all about* Sagan. That's a very dangerous place to be for a scientist. Vested interests love to single out an individual scientist for attack (like they did with me and the hockey stick curve), making an example of them for other scientists who might consider stepping forward. I've called it the "Serengeti Strategy." Sagan had unfortunately made it easy.[27]

People's personal feelings about Sagan influenced how they viewed the nuclear winter threat. He was outspoken—a rare quality in scientists at the time—and so easy to attack. The columnist William F. Buckley Jr. said Sagan was "so arrogant he might have been confused with, well, me."[28]

Any internal scientific differences or disputes were greatly amplified in the larger public debate over nuclear disarmament, particularly by ultraconservative Cold War physicists. Especially influential was S. Fred Singer. Singer was an all-purpose, industry-funded purveyor of denial. In 1990, he left a faculty position in the Department of Environmental Sciences of the University of Virginia to form his own organization, funded by big tobacco, fossil fuel companies, and other corporate interests, with the Orwellian name Science and Environmental Policy Project (SEPP). Singer used the SEPP as a platform for advocating against what he called the "junk science" of acid rain, ozone depletion, tobacco health threats, and climate change. I was at the receiving end of many of his critiques and attacks over the years.[29]

Singer was perhaps the most vociferous critic of Sagan and nuclear winter, dismissing the work as sensationalistic and based on fundamentally flawed modeling. In 1983, he wrote that "Sagan's scenario may well be correct, but the range of uncertainty is so great that the prediction is not particularly useful." Note the way that scientific uncertainty is cited as an argument for inaction, when the principle of precaution suggests just the opposite. The uncertainty trope would be recycled many times over by others seeking to undermine environmental policy, be it acid rain, ozone depletion, or climate change. Singer engaged in a full court press criticizing Sagan and the science of nuclear winter in the ultraconservative editorial pages of the *Wall Street Journal*. He also managed to get letters to the editor critical of nuclear winter published in the two leading scientific journals, *Nature* and *Science*, the latter being the journal that published the original TTAPS article.[30]

Singer was just the tip of the spear, however. Let's talk about the Strategic Defense Initiative, or SDI, proposed by the Reagan administration in March 1983, with the avid support of Cold War hawks and military contractors. SDI was a hypothetical space-based anti-missile defense system aimed at intercepting Soviet missiles with space lasers. Given that it sounds like the stuff of science fiction, it's hardly surprising it became known as "Star Wars" after the very popular movie trilogy. Sagan actively campaigned against SDI, arguing that it would lead to a further escalation of tensions between the United States and the Soviet Union, a dangerous buildup in nuclear arms, and a heightened threat of a catastrophic nuclear winter scenario. Now he was taking on the entire military-industrial complex. And they weren't going to take to that kindly.

A group of three physicists who had cut their teeth on the Cold War weapons program joined forces to attack Sagan and the science of nuclear winter, coming together to form the George C. Marshall Institute (GMI)—a conservative think tank whose primary purpose was to oppose Sagan and others warning of the threat of nuclear winter and to provide support and advocacy for SDI. The head of the group was Frederick Seitz, a solid state physicist, former head of the U.S. National Academy of Sciences, and winner of the Presidential Medal of Science. The group also included Robert Jastrow, founder of the NASA GISS laboratory later led by James Hansen. Jastrow—a skilled communicator in his own right—had been nudged off *The Tonight Show* stage by the more charismatic Sagan. It's easy to imagine that there might have been some residual bitterness on his part. The third member was Nicolas Nierenberg, onetime director of the Scripps Institution of Oceanography—today a leading climate science research institution.[31]

The GMI trio saw legitimate concerns about the SDI program as scare tactics employed by Soviet-sympathizing peaceniks. They viewed the concept of nuclear winter as a threat to our security. And so they teamed up with conservative politicians, industry special interests, and right-wing media in an effort to undermine public faith in the underlying science—first by discrediting Sagan, personally. Their attacks took the form of congressional briefings and the writing and placement of popular articles and op-eds aimed at debunking Sagan's science. It even included intimidating television networks when they considered running a nuclear winter documentary.[32]

Just as this was all playing out, I happened to be doing my undergraduate degree in the very epicenter of the academic debate over SDI: the UC Berkeley Physics Department. The physics faculty included Edward Teller, who was also a cofounder of the Lawrence Livermore Laboratory, a nuclear weapons laboratory that operated under the auspices of UC Berkeley. He is often called the "father of the hydrogen bomb" for his central role in the Manhattan Project in the 1940s. Teller was afforded an honor that was otherwise granted only to Nobel Prize winners at Berkeley: a reserved parking spot on campus, right next to LeConte Hall (the physics building), with his name on it.

Teller was one of the primary advocates of SDI and, unsurprisingly, one of Sagan's fiercest critics. Teller wrote to Sagan, "My concern is that many uncertainties remain and that these uncertainties are sufficiently large as to cast doubt on whether the nuclear winter will actually occur." He even turned to name-calling: "I can compliment you on being, indeed, an excellent propagandist—remembering that a propagandist is the better the less he appears to be one." Teller attacked Sagan's work in a 1984 commentary in the journal *Nature*, the summary paragraph of which read: "Today, 'nuclear winter' is claimed to have apocalyptic effects. Uncertainties in massive smoke production and in meteorological phenomena give reason to doubt this conclusion." Teller added: "Highly speculative theories of worldwide destruction—even the end of life on Earth—used as a call for a particular kind of political action serve neither the good reputation of science nor dispassionate political thought."[33]

It was one thing to disagree with Sagan's science. But it was something else entirely to question his motives, objectivity, and honesty. I frankly find it shocking that this ad hominem language made it past the editors of one of our two leading scientific journals. It is perhaps relevant, in this regard, that the editor-in-chief at *Nature* at the time, John Maddox—a theoretical physicist by training—was himself a nuclear winter critic. He published nearly a dozen articles, commentaries, and editorials in *Nature* criticizing Sagan and nuclear winter in the two years following the publication of TTAPS in *Science*, including two very critical editorials of his own. The fact that our oldest, arguably most-revered scientific journal had effectively been hijacked in the effort to discredit Sagan personally inevitably feeds the notion of a politically and ideologically motivated pile-on, with even some of our most celebrated scientific institutions bearing some culpability.

While Teller was a member of the UC Berkeley physics faculty, so, too, was one of the most ardent critics of SDI. Charles Schwartz helped spearhead the anti-SDI pledge, signed by almost 7000 scientists and engineers, calling for a boycott of SDI research. I recall a friend of mine who was taking a physics class with Schwartz at the time relating, with some combination of amusement and disbelief, how Schwartz had canceled class one day so that students might participate in an SDI protest taking place in Sproul Plaza—the site of the famous Berkeley demonstrations in the 1960s and 1970s. By 1986, Schwartz had refused to teach any more courses to physics majors, in protest of the involvement of the physics community in furthering what he saw as a misguided and dangerous gambit.[34]

I must confess that I was largely disengaged from this debate as a Berkeley physics major in the 1980s. I was focused on physics and math problem sets, studying for exams, and my initial forays into physics research. I was oblivious to the fact that our department was home to prominent adherents to polar opposite positions in the defining socio-scientific debate of our time. I look back with some astonishment at what was actually going on around me, and the larger political and ideological battles that laid behind it. And I ponder with some fascination the fact that the department was *also* home to physicist Luis Alvarez, who first identified the ancient asteroid impact behind the deep freeze that killed the dinosaurs and, in many respects, raised the specter of nuclear winter in the first place. Synchronicity rears its head again?

Where this episode becomes even more germane to the central topic of this book is that the nuclear winter simulations by Sagan and colleagues were based on early-generation global climate models. That put climate modeling firmly in the sights of the GMI trio, Fred Singer, and other like-minded contrarians, and it set the stage for their subsequent role in the fossil-fuel-industry-funded assault on the science of climate change. With the collapse of the Cold War in the late 1980s, the GMI gang needed another issue to focus on. Denial of acid rain and ozone depletion would keep them busy through the early 1990s. But as these matters faded from view, in part because there was bipartisan support for policies to deal with them, GMI had to find some other justification for their continued existence. Climate change denialism fit the bill, and GMI—to quote *Newsweek*—became a "central cog in the [climate change] denial machine."[35]

Though climate change has eclipsed nuclear winter as a concern, that doesn't mean that it's gone away as a threat. Rutgers nuclear winter expert Alan Robock, whom we encountered earlier, has continued to explore nuclear war scenarios with far more sophisticated climate models than those used by Sagan, Schneider, and others back in the 1980s, reinforcing many of their main findings. The models used today by Robock and collaborators capture vertical motion in the atmosphere far more accurately, demonstrating that smoke particles can be lifted into the upper stratosphere, where they can remain for many years. That extends the lifetime of their cooling influence. Renewed Russian aggression reminds us that the Cold War isn't over, and as Robock notes, there are now nine countries with nuclear weapons. A nuclear conflict between India and Pakistan, for example, could still have catastrophic consequences.[36]

Strangelove Returns

Let us return to the discussion of the K-Pg extinction event that began
this chapter. Despite the overwhelming evidence that the mass extinction was caused by a deep freeze induced by the collision of Earth with
a giant asteroid, some contrarians—curiously, often climate change
deniers—have continued to contest that interpretation for decades.
Telling in this regard is an exchange some years ago between climate
contrarian Fred Singer and mainstream climate scientist Alan Robock, both of whom we encountered earlier in this chapter. During a
debate in 1997 about human-caused climate change, the cause of the
K-Pg event came up at one point. Singer insisted that the dinosaurs
"actually did very well until they were hit by an asteroid." Robock
gently corrected him: "Actually, they weren't hit by the asteroid. It
was the climate changes induced by the asteroid that changed their
environment so much." Robock is correct, of course, but one can
understand why an admission of that fact would be inconvenient to
a climate change denier like Singer.[37]

Some of the contrarianism, however, is simply scientific skepticism playing out as it is supposed to. True scientific skepticism—
as opposed to politically motivated denialism—is, after all, part of
what Carl Sagan called the "self-correcting machinery" of science.
Some geologists, for example, have implicated volcanic activity in
an igneous province (located in India) called the Deccan Traps that
might have injected substantial amounts of CO_2 into the atmosphere
in a relatively short amount of time. Recent work, however, seems to
put to rest this notion. One study tested the viability of both Deccan
eruptions and the Chicxulub asteroid impact–induced cooling, concluding that only the latter fits the extinction pattern. Other studies
conclude that the asteroid impact might actually have triggered the
Deccan Traps eruptions, which could have even been a *mitigating*
factor, offsetting some of the cooling and helping some species avoid
extinction.[38]

Just as the overwhelming majority of scientists now accept the
science of human-caused climate change, so, too, do they accept that
the K-Pg mass extinction was caused by a massive impact event.
And that was before the recent nail in the impact-hypothesis-denial

coffin: direct evidence we encountered earlier in this chapter of an impact-generated flood along with the preserved remains of dinosaurs mixed together with pieces of the asteroid itself.

That doesn't mean that other factors didn't come into play. Walter Alvarez, coauthor of the original 1980 asteroid impact study, acknowledged as much years ago. One recent study, using a careful statistical approach to assess both speciation (the generation of new species) and extinction (the loss of existing ones) over time, concluded that there was long-term decline over the preceding ten million years in the ability of dinosaurs, as a group, to replace extinct species with new ones, implying increased vulnerability that might have made them especially susceptible to extinction when the asteroid struck.[39]

The pattern of extinction in the ocean and the nature of the post-impact recovery has remained somewhat of a mystery and continues to be debated by scientists. And it leads us back to the concept of the Strangelove Ocean. We first encountered the term to describe an ocean that became nearly devoid of life (ninety-six percent extinction rate) at the end of the Permian in response to the multiple insults of dramatic warming, deoxidation, acidification, and a toxic sulfur dioxide "stink bomb" thrown in for good measure.

The term, however, actually originated in 1982, in an article by prominent Columbia University geologist Wallace Broecker that described a similar inferred ocean die-off in the wake of the K-Pg impact event. The Strangelove reference is seen to make a whole lot more sense in this context. The film *Dr. Strangelove*, after all, satirized nuclear conflict between the Soviet Union and the United States. These thematic connections between the K-Pg asteroid impact and the Cold War would have been top-of-mind to any Earth scientist studying the K-Pg event in 1982, even if four decades later the film reference might be lost on some.[40]

In some sense, the K-Pg event seems like a less dramatic example of ocean die-off than the P-T event. The rate of extinction of ocean organisms was substantially lower (roughly seventy-five percent). Nonetheless, there does appear to have been a remarkable drop in ocean productivity. To understand just what happened, we'll need—I apologize—to talk once again about carbon isotopes.

We can assess the history of ocean productivity by looking at carbon isotopes in deep-sea sediments. Forams are microscopic marine organisms that form a calcareous shell. When they die, their shells sink to the ocean floor, becoming part of the sedimentary record. By drilling a core through the sediment at the ocean bottom, scientists can recover a chronology back in time of these organisms. Different types of forams are known to live at different levels in the ocean. Those that live somewhere in the water column are called planktic, and those that live down in the seabed are known as benthic. Photosynthesizing organisms, as we learned in the previous chapter, prefer to take up carbon-12 during photosynthesis, which leaves a greater relative abundance of carbon-13 in the water column, with which calcareous marine organisms build their shells. The difference between carbon-13 and carbon-12 (called delta C-13) in foram shells is a measure of ocean biological productivity, and the difference in delta C-13 between shells of planktic and benthic species is a measure of the burial of organic carbon in the deep ocean.

Sediment cores recovered from the deep ocean that contain the K-Pg transition show a prominent decrease in this difference immediately after the K-Pg boundary, implying a sharp drop in organic carbon burial. This observation is the basis of the original Strangelove Ocean model of an essentially lifeless ocean. There are a couple problems with that model, however. First of all, the decreased organic carbon burial persisted for around three million years, far exceeding the duration of any direct effect of the impact event on sunlight or climate. Second, though planktic species display a sharp decrease in delta C-13, benthic species do not. That is inconsistent with a "dead ocean."[41]

These observations have motivated the alternative Living Ocean model, which posits that the K-Pg event altered the ecology of the ocean by rendering extinct the larger, more vulnerable ocean biota, making it more difficult for organic matter to sink to the ocean floor, because smaller particulates are more easily suspended in the upper water column, whereas larger particulates sink. That, in turn, would have reduced carbon burial. In this model, ocean productivity continued on, but with a much-reduced flux of carbon to the deep ocean.

The original proponent was my collaborator, University of Rhode Island oceanographer Steven D'Hondt.[42]

Working with D'Hondt and me back in 2004, a former graduate student of mine, Brad Adams, analyzed carbon isotope data derived from deep-sea sediment cores from both the Atlantic and Pacific Ocean basins that span the K-Pg boundary to statistically characterize what happened. Our results indicated a two-stage recovery. There was an initial gradual recovery over the first three million years to an intermediate state of carbon burial, followed by a step-like return to near pre-impact levels by four million years or so after the impact event. The pattern and timing of the carbon cycle recovery seems to suggest the role of key biological events. Though the direct climate impact of the asteroid strike would have dissipated in a matter of decades, it took several millions of years for critical innovations, like the evolution of new, larger calcareous ocean biota, to fill the niches of those that went extinct. The ocean had to redevelop the entire trophic structure and food web ecology, and that takes time—time measured not in years but in millions of years.[43]

I was able to see one of the actual sediment cores we had analyzed eight years later. It was October 2012 and I was speaking at the South by Southwest (SXSW) "Eco" conference in Austin, Texas, about my latest book, *The Hockey Stick and the Climate Wars*. Following the conference, I drove a hundred miles to College Station, Texas, for a lecture and visit at Texas A&M University, which included a tour of the international ocean drilling program repository that is sited there. The drive took me right through the charred remains of the Bastrop County Complex fire, the most destructive wildfire in Texas history. The fire began on September 4, 2011, following a summer of unprecedented heat and drought. It burned for fifty-five days, engulfing 32,000 acres. The loblolly pine forest that was destroyed was an example of what is known as a relict forest—a forest that won't grow back in today's hotter and drier climate. It was a sober example of tipping points and the phenomenon of hysteresis—a reminder that some things are lost forever. There is no going back. I imagined what it was like when this ancient forest was engulfed in flames, flames perhaps not unlike those that burned the forests of the world in the wake of the K-Pg impact event, or those that might incinerate entire cities in the event of a global nuclear war.

Other recent work seems to confirm our earlier findings of a staged recovery. A 2019 study by Michael Henehan of the Yale Department of Geology and Geophysics determined that the oceans experienced as much as a fifty percent loss in biological productivity in the immediate aftermath of the impact event, followed by a transitional period in which marine productivity recovered. According to Henehan, "In a way, we reconciled both of these 'Strangelove' and 'Living Ocean' scenarios. Both of them were partially right; they just happened in sequence."[44]

We know that the asteroid struck an area of carbonate rock containing a large amount of sulfur, which would have vaporized into sulfur dioxide upon impact, creating reflective sulfate aerosols. The aerosols would not only have added to the cooling effect of the other particulate matter ejected into the atmosphere by the collision, but it would also have caused widespread acid rain and ocean acidification. Henehan and colleagues made use of boron isotopes to assess the drop in ocean pH following the K-Pg impact event, finding evidence for a substantial increase in ocean acidity, which would explain the selective extinction of calcareous ocean biota and the drop in deep-sea carbon burial in the wake of the impact.[45]

Lest this all seem like a purely academic debate, it's worth noting that this scientific tale offers some important lessons for our predicament today. Though surface warming will stabilize relatively quickly after human carbon emissions go to zero, ocean acidification will continue on for centuries, and it will continue to pose a threat. One thing we learn from the post K-Pg recovery is that a massive die-off can lead to the collapse of food chains and disruptions of ocean trophic structure that take not years or decades, or even centuries or millennia, but millions of years to recover. Even the notion of full recovery is questionable, of course, for many key species are lost permanently. Though their niches may be filled once again, they will never be brought back. It's something to ponder as we continue to engage in an unprecedented disruption of our planetary environment.

Lessons in Our Past

If you're looking for a ray of hope, you could, I suppose, take some limited solace from the fact that the dinosaurs didn't *quite* die out after all. As I wrote this paragraph, I was watching a pair of them through the window looking out on our back yard: two blue jays feasting on the budding early autumn fruit of our crabapple tree. Modern birds descended from a group of two-legged, three-toed, hollow-boned dinosaurs known as theropods, whose members include the imposing *Tyrannosaurus rex*. But the group also included much smaller species that, like small mammals, were better equipped to deal with the sudden freeze. They gave rise to birds.

But that's really a technicality. A better reason for optimism is this essential distinction: there was nothing the dinosaurs could have done about their plight. They had no means to deflect the asteroid. They lacked *agency*. We do not. We are threatened with a catastrophe of our own making. And the primary challenge we face isn't the immutable laws of astrophysics. It's political will.

As the threat of nuclear conflict mounted in the 1980s, two leading scientists representing the two opposing nations—Carl Sagan in the United States and Andrei Sakharov in the Soviet Union—joined together in a common effort to communicate the existential threat of a nuclear conflict to their respective people and to convince their heads of state that global nuclear war was unwinnable and the global arms race dangerous and misguided. That effort proved successful. In December 1987, Ronald Reagan and Mikhail Gorbachev signed the Intermediate-Range Nuclear Forces (INF) Treaty, banning all short and intermediate range ballistic missiles, heralding the ostensible end of the Cold War. Both had cited concerns about nuclear winter.[46]

Over the past two years we have, unfortunately, seen a renewal of Cold War tensions in the wake of Russia's war on Ukraine. Russian president Vladimir Putin has hinted at the use of tactical nuclear arms in the escalating conflict. We have also seen Russia emerge as a petrostate, increasingly dependent on the monetization of their fossil fuel assets. Putin has engaged in ever-more audacious efforts to block global climate action. It is arguably more important than ever that scientists follow the lead of Sagan and Sakharov and engage in joint efforts that transcend national boundaries to advance the cause of science-based policy. Such was indeed the call to arms issued in a recent commentary by several of my colleagues in *The Hill* that warned that the "disruption of relations among the great powers makes it even harder to sustain the international collaboration needed to tackle climate change. Its chilling effect extends not only to cooperative efforts to meet global emissions goals, but also to the research and policy studies that are needed to guide global action."[47]

Here is another lesson that mighty *Brontosaurus* has for us: there are always winners and losers. Though the (non-avian) dinosaurs went extinct, small, shrew-sized mammals were the big winners. They are our mammalian ancestors. Without their predators around, small mammals were safe to come out of their holes and cracks, and they would eventually flourish and fill the various new niches that emerged. If we extinguish ourselves, other creatures will undoubtedly exploit the niche we had filled. They'll be the winners. And we'll be the losers. Yes, the planet itself will continue on just fine. But without us. *Our* fragile moment will be over.

Does climate change indeed threaten *our* extinction? In seeking an answer to that question, we will look to one more event in our deep past—the one that comes the closest to being a possible analog for catastrophic human-caused warming: the so-called PETM. Global temperatures then might have reached a sauna-like 90°F. This extreme warmth may have triggered a major release of methane and perhaps other Hothouse Earth–amplifying feedback mechanisms. Could that conceivably be in store for us if we continue to burn fossil fuels with abandon? We address that question next.[48]

5

Hothouse Earth

How can anyone survive in a climate like this? A heat
wave all year long.

—*Soylent Green*

A t the opposite end of the spectrum of Snowball Earth is Hothouse Earth. As we've seen, vicious cycles of amplifying feedback processes can lead to a runaway cooling of Earth given enough
initial cooling. But the same processes, like the rapid melting of ice,
can amplify warming once it starts. And there are other amplifying
feedbacks that kick in when things get *really* warm. Could feedback
processes involving the release of methane or the behavior of water
vapor and clouds ratchet up the warming at high greenhouse gas levels, ending our fragile moment? Geological evidence from past hothouse climates, as explored in this chapter, can potentially provide
the answers we seek.

The PETM

The poster child for Hothouse Earth is the so-called PETM, short
for the Paleocene-Eocene Thermal Maximum. The PETM was an
episode of rapid (geologically speaking) warming that occurred fifty-
five million years ago, just ten million years after the demise of the
dinosaurs. What is unique about the PETM among other episodes of

rapid planetary warming in our geological past is the amount of carbon that was released into the atmosphere in a short time period, and the rate of the warming that resulted. It's probably the best natural analog we have for the current human-caused spike in CO_2 levels and global temperature.

The warming spike occurred atop a more gradual warming trend that had begun in the early part of the Paleocene (the epoch that began with the K-Pg impact) and had continued into the beginning of the subsequent Eocene. During the PETM, average global temperatures increased by approximately 9°F, starting from a baseline that was already about 18°F warmer than today. The bulk of the warming occurred in as little as 10,000 years. That's about 0.05°C (0.09° F) per century, small compared to the current rate of roughly 1°C (1.8° F) per century, but extremely rapid by geological standards. Though the input of carbon that triggered the PETM happened over just thousands of years, the elevated warmth persisted for 200,000 years after that. That's an important lesson for us about our troublesome legacy given that we're adding carbon to the atmosphere even more rapidly today through fossil fuel burning.[1]

How do we know how much the planet warmed and how quickly? As we saw in Chapter 3 with the Permian-Triassic (P-T) mass extinction, it is possible to estimate ancient temperatures from the ratio of stable oxygen isotopes (oxygen-16 and oxygen-18) in the preserved shells of calcareous biota. But that ratio *also* depends on global ice mass (the light isotope ends up getting trapped in continental ice, leaving more heavy oxygen in the ocean). Because there was no ice during this period, we can eliminate that factor. However, *salinity* changes also influence the isotope ratios (there are other technical complications due to potential for alterations in carbonate chemistry over time). Several other lines of evidence nonetheless help us estimate past ocean temperature changes. They include "clumped" isotopes, which analyze carbon-oxygen bonds, and ratios of seawater magnesium and calcium incorporated into calcite shells, which are tied to the temperature of the surrounding seawater. There's also the more complicated paleothermometer based on lipids (fatty layers) from microbes in sediments. Though there are limitations and caveats with each of these data sources, collectively they suggest that the surface ocean, the deep ocean, and global average surface temperatures all warmed between 7°F and 11°F over the course of the PETM.[2]

We also know that there was a massive addition of inorganic carbon into the ocean-atmosphere system, thanks to stable carbon isotopes, which show a large downward spike in delta C-13, though just how much is rather uncertain. One complication is that the carbon could have come from both carbon dioxide (CO_2) and methane (CH_4), each of which has very different warming potential. And there are multiple sources of both, each of which has a different delta C-13 signature.

In a 2011 study, researchers used a simple model known as a box model to distinguish between the different sources of carbon dioxide and methane, finding that the carbon input most likely occurred in two distinct stages, including an initial release of about 1000 gigatons of carbon (GtC) tied to the first 3000 years of PETM warming, followed by a rapid input of about 1200 GtC tied to the big negative isotopic spike and to an additional 1000 years of warming. They concluded that the carbon was released over a timeframe of less than 500 years. Though they could not pinpoint the cause of the initial release, they concluded that the second input of carbon, which yielded a negative spike that bears the very negative isotopic fingerprint of methane, was likely a crystalline form of methane known as methane hydrate, trapped in a "cage" of water molecules that is found in sediments along continental margins (and in permafrost). Methane hydrate can be destabilized by ocean warming, constituting an important potential amplifying carbon cycle feedback.[3]

A more recent 2016 study used a combination of oxygen and carbon isotopic data to try to further pin down the amounts of both CO_2 and CH_4 that were released during the PETM. Though the isotopic analyses we've focused on thus far involve the two main stable isotopes oxygen-16 and oxygen-18, it turns out that you can tease out some more information if you add oxygen-17 to the mix. To do so, the authors turned to an unlikely source: the tooth enamel of ancient, preserved mammals (yes, scientists in the future might look at your teeth to piece together the puzzle of how we warmed up the planet). The levels of the three different oxygen isotopes in the water that makes up mammalian bodies are determined by a number of factors, but the oxygen-17 specifically scales with the level of CO_2 in the atmosphere. In other words, it can be used as a proxy measure of ancient CO_2 levels. Knowing something about those levels helps you tease apart the competing influences of CO_2 and CH_4 on carbon isotope ratios. These scientists concluded that while CO_2 levels rose, it was to no more than about 2500 ppm in the atmosphere. That means that a substantial release of methane was required to explain the PETM delta C-13 spike.[4]

Why should we worry so much about a huge input of methane? After all, as we learned earlier, its life span in the atmosphere today is short compared to that of CO_2—decades rather than millennia, thanks to oxidation in Earth's oxygen-rich atmosphere. That's largely been true since the rise of oxygen-generating photosynthesizing life more than two billion years ago. That doesn't mean, however, that it has no long-lasting impact. When methane oxidizes, it turns into CO_2. We get one CO_2 molecule for every methane molecule released. So a huge amount, like 1000 GtC, of methane gives us a huge amount—1000 GtC—of CO_2. That means that the long-term warming impact of carbon emissions is dictated by the total slug of GtC released, regardless of how the initial input of carbon was partitioned between carbon dioxide and methane.

We now have an idea of how much warming there was, and at least some vague idea of much carbon might have been released. How can we pin down the timing? Unlike other archives we'll encounter for later periods, like corals, ice cores, or tree rings, we can't build a reliable year-by-year chronology. Dating ancient sediments is like dating a *flake*—it comes with substantial uncertainty. Age estimates can be off by many thousands of years. To try to get around that, Richard Zeebe of the University of Hawaii and collaborators employed a novel modeling approach. They exploited the *relative* lag time between carbon isotope ratios (which record the carbon release) and oxygen isotope ratios (which track the climate response). They concluded that the maximum rate of carbon emission during the beginning of the PETM was just over a billion tons (a gigaton) of carbon a year, which lasted more than 4000 years.[5]

Within the broad uncertainties, and sometimes conflicting estimates, derived from different data sources and different modeling
approaches, we can reasonably conclude the following: the planet
warmed up between 7°F and 11°F during the PETM due to a large
release of carbon, somewhere between 2000 and 15,000 GtC. We
know that the carbon was released over a timeframe of between 2000
and 50,000 years (a large uncertainty range for sure—but that's what
you're trying to reconstruct an event that happened
more than fifty million years ago).

Current estimates are that fossil fuel reserves contain between
1000 and 2000 GtC and potentially as much as 13,000 GtC if we
were to mine and burn all accessible oil, gas, and coal. That's about
as much carbon as was released during the PETM, but over a timeframe of centuries rather than millennia. It would likely be enough
carbon pollution to melt the entire Antarctic Ice Sheet, raising global
sea level by 160–200 feet, submerging highly populated areas that
are home to more than a billion people, including New York City and
Washington, DC.[6]

Enter the Inferno

What was the PETM actually like? Of course, we don't have photos
or documentary evidence, but a rather striking picture emerges from
the fossil record of plants and animals. Mangroves and rainforests
reached Arctic latitudes. Hippos, alligators, and palm trees graced
Ellesmere Island, off the northwestern coast of Greenland, suggesting
lush, balmy conditions near the North Pole. There's evidence that
some tropical ocean regions became so hot that they were abandoned
by many organisms.[7]

My former Penn State colleague and friend Timothy Bralower is
one of the leading experts in the world when it comes to the PETM.
He notes that temperatures were a balmy 68°F off the coast of Antarctica, where it is close to freezing today, and a scorching 97°F off
the coast of West Africa, adding, "I've been swimming in Miami
in August and it feels like a bathtub at 88°F, but 97°F is virtually
uninhabitable!"[8]

The Bighorn Basin in Wyoming today is home to the badlands, a dusty northern desert environment of scrub and sagebrush. During the late Paleocene, just prior to the PETM, it was subtropical forest, similar to northern Florida today, with swamps of bald cypress, palm trees, and crocodiles. During the PETM, mean annual temperatures appear to have reached 79°F, more similar to southern Florida. The swamps disappeared and rainfall turned more intermittent. It became hotter but also drier.[9]

Was the drying part of a widespread trend? Probably not. Fossil pollen evidence suggests that tropical forests flourished and spread at this time. Climate model simulations of the PETM, using elevated CO_2 levels consistent with the paleo data, suggest that western North America was likely one of the exceptions to the rule, one of the handful of continental mid-latitude regions that saw drying, primarily during summer, due to high surface pressure and the poleward migration of the jet stream. Many other regions—particularly in the tropics and subpolar latitudes—likely saw increased precipitation. Warmer air holds more moisture, so when conditions are favorable for rainfall—which they would have been over much of the planet—you get even more of it.[10]

Over a large part of the planet, it would have been both very hot and very humid. That's a bad combination. The old cliché is that "it's not the heat, it's the humidity," but as anyone who has been to Las Vegas in August will tell you, that's not true. It's both. In fact, the best measure of susceptibility to heat stress combines temperature and humidity into a single variable. It is called the wet bulb temperature.

When I was a graduate student at Yale, I was a teaching assistant for the very popular undergraduate course "Oceans and Atmospheres" taught by my former Ph.D. committee member Ronald Smith. One of my favorite labs was the weather lab, where we would take the students up on the roof of the Kline Geology Laboratory and introduce them to a standard meteorological station, containing various types of weather instruments inside a little white shed elevated a few feet off the ground. Among the more interesting instruments was the sling psychrometer. It consisted of two thermometers attached side by side. One had a cloth wick covering on the bulb, which you saturate with water. There's also a little rope attached to the instrument that allows you to "sling" it around in the air, speeding up the evaporation of the wet bulb thermometer, which cools off in response to the heat lost due to evaporation. It eventually reaches some new, lower equilibrium temperature.

The difference between the dry bulb (no cloth covering) and wet bulb thermometer readings is a measure of the relative humidity of the atmosphere on that day. And the wet bulb temperature measures the lowest temperature an object—which could for example be a human being like you or me—can reach through evaporative cooling at the prevailing temperature and humidity.

Our core body temperature is typically about 98.6° F. Mine runs somewhat lower, about 97°F. That makes me a veritable cold-blooded reptile compared to my wife and daughter, a source of endless battles over the setting of the thermostat. Skin temperature is typically lower by 4–9°F, depending on the level of physical activity, which helps transfer excess heat from the body's core to the skin and then the surrounding air. Sweating helps keep the core temperature from rising, but it becomes increasingly ineffective as a cooling mechanism the more humid the air. A wet bulb reading of 86°F exceeds guidelines for safe physical activity, and a wet bulb temperature of 90°F, which feels as hot as a dry temperature of 131° F, is dangerous even without physical activity. Wet bulb temperatures of 95°F are comparable to a dry temperature of 160°F. At this point, your skin can no longer shed excess heat to the air. Even in the shade you will die in a matter of hours.[11]

In most places today the wet bulb temperature never exceeds 86°F. The 95°F wet bulb survivability limit, however, has now been exceeded at least briefly in some locations in South Asia, the coastal Middle East, and coastal southwest North America. These are regions with close proximity to both very high sea temperatures and extreme summer heat—conditions jointly conducive to exceptionally high wet bulb temperatures. In *The Ministry for the Future,* science fiction writer Kim Stanley Robinson begins his engaging fictional account of our near-term climate crisis future with a heat wave in India where wet bulb temperatures remain above 95°F for days on end, killing twenty million people. Life, alas, is beginning now to imitate art. During a historic early-season heat wave in spring 2022,

temperatures in Chennai, India, reached 94°F with seventy-three percent relative humidity. That's a wet bulb temperature of 86°F. And
that was in early May.[12]

With as little as 4.5°F additional global warming, something we
could witness in a matter of decades in the absence of substantial climate action, this limit could be exceeded regularly in parts of South
Asia and the Middle East that are home to as many as three billion
people. Warming of 18°F could eventually be reached if we burn all
estimated fossil fuel reserves (or if we burn a fair share of them and we
happen to encounter some nasty destabilizing feedbacks). At that level
of warming, much of the human population could at least occasionally
be subject to this deadly heat limit. "A heat wave all year long" indeed.[13]

Some of my own research involves looking at climate model projections to assess the potential for severe heat exposure in the United
States. My collaborators and I recently examined simulations used
by the Intergovernmental Panel on Climate Change (IPCC) to project
future changes in heat stress accounting for both heat and humidity. We found that short- to medium-duration episodes of extreme
heat stress are likely to increase more than three-fold across densely
populated regions of the United States in the Northeast, Southeast,
Midwest, and desert Southwest by the end of this century in a high
carbon emission scenario. Other research I've conducted suggests
that these very same model projections may be underestimating the
true heat stress risk, as the models tend to underestimate the wavy
summer jet stream conditions that are associated with the most persistent heat extremes. Adverse heat-related health impacts are already
being experienced by outdoor workers in Las Vegas, Los Angeles,
and Phoenix.[14]

It is relevant, in this context, to imagine what conditions might
have been like during the PETM when the global average temperature was an almost inconceivably hot 90°F. That's 30°F warmer
than today. Adding insult to injury, it was generally humid as well.
What would it be like to suddenly be transported back in time to the
PETM? There's a good chance you could find yourself experiencing
a daytime high temperature of 100°F and eighty-two percent relative
humidity. That combines the heat of a sauna with the humidity of
a steam room, a hybrid entity that humans don't build in the real

world for good reason—it's deadly. Eighty-two percent relative humidity and 100°F amount to a wet bulb temperature of 95°F. If you found yourself subject to those conditions without access to refrigeration, air conditioning, or a cold pool to dive into, you'd soon die from exposure.

The conclusion is that *Homo sapiens*—at least in our current form, and without the luxury of modern technology—couldn't have lived during the PETM, except possibly near the poles. Yet, many other mammals appear to have made out fine. There was a die-off of deep ocean biota—likely related to the anoxic conditions induced by deep ocean warming. But we didn't see any mass extinction of mammals. Instead, we saw a combination of migration and evolution. Evolution works if you've got thousands of years to work with, which is a luxury we don't have today, but which PETM life-forms did. The primary adaptation, other than migrating poleward in an effort to escape the heat, was *dwarfing*.[15]

As a rule, the way to cool down is to get smaller. To invoke a shopworn science-nerd joke that teases physicists for the sometimes laughably unrealistic simplifications they make in modeling the real world, consider a spherical cow. Only I'm not kidding. Imagine that a cow was indeed spherical. The volume of a sphere, as you might remember from high school geometry, is $4\pi R^3/3$ where R is the radius.

Figure 13. A spherical cow.

The surface area of that same sphere, you might also recall, is $4\pi R^2$. So, the surface-to-volume ratio is $3/R$. It decreases with the radius. The larger the (spherical) cow, the smaller its surface-to-volume ratio. Now think about what happens when the cow heats up, that is, when the entire body, which is the spherical "volume," warms up. The cow can only cool off by losing heat through its skin, or through its "surface." The larger it is, the smaller its surface-to-volume ratio and the harder it is for the cow to cool off.

I know what you're thinking at this point: Cows aren't spherical. They're *cubic*! Well, the argument still applies. If L is the side length of the cubic cow, the surface area is $6L^2$ (the sum of the areas of all six square faces, each of which has area L^2), whereas the volume is just L^3. So, the surface-to-volume ratio is $6/L$. The larger the cow, the smaller the surface-to-volume ratio. As it happens, this rule applies even to an arbitrarily shaped cow, or any animal at all. There's even a name for it: Bergmann's rule. It says, in short, that the best way to cool down is to get smaller. That's what happened during the PETM—in an evolutionary sense. Larger representatives of a given mammal species couldn't get rid of heat very well. They selectively died off and failed to reproduce. The smaller members of their species were more likely to survive and more likely to pass along their genetic traits. And so on. That's how the process of dwarfing proceeds.

We see some dramatic examples during the PETM. Horses, which had only recently appeared on the scene, shrunk by thirty percent in size (and scaled back up seventy-six percent as the PETM came to an end). It would be tempting to conclude that the warming therefore wasn't a big deal. They just adapted, after all. Much as some critics of climate action insist that *we* will just "adapt" to the impacts of climate change. But understand that any selective pressure so great as to shrink animals by thirty percent in over as little as 10,000 years, implies substantial mortality of those with maladaptive traits (namely, larger size). Think about that fact the next time you hear a climate contrarian insist we can simply "adapt" to climate change. It is true that our *species* can likely survive 9°F warming. It is also true that hundreds of millions of our fellow human beings will likely perish from it.[16]

As always, there were winners and losers in the PETM. Our primate ancestors were winners. Even though the PETM would have been too hot for us, it did provide a selective advantage to our much smaller progenitors, the first primates. The initial representative was the primitive, lemur-like, mouse-sized, arboreal, vegetarian *Dryomomys* we learned of in Chapter 1.

Ocean bottom–dwellers, on the other hand, were the losers. While surface forams and upper ocean biota in general made out okay, deep-sea benthic forams were devastated by acidification. Perhaps as much as fifty percent of all benthic foram species went extinct. In fact, deep-sea acidification was so extreme that sediment cores are relatively devoid of calcite shells, many of which were literally dissolved. And we think that lesser, but still significant, acidification took place in the upper ocean.[17]

Ocean circulation changes, too, might have played an important role in the PETM. A clay mineral known as kaolinite is produced by silicate weathering and carried off in streams and rivers to the ocean. The fact that anomalous levels of kaolinite are found in PETM-dated ocean sediments suggests that the overall increase in rainfall during the PETM likely led to increased continental runoff, delivering large amounts of freshwater to the ocean. As we've seen before, a large input of freshwater to the ocean can disrupt the so-called ocean conveyor belt circulation. A combination of climate model simulations and carbon isotope data from forams shows that not only did this disruption occur, but that it led to the burial of warm, oxygen-depleted waters at the ocean bottom. The acidification, warming, and deoxygenation of the deep ocean would have constituted a triple whammy for deep-sea life.[18]

The deep ocean warming, estimated to be as much as 5–7°F, could have destabilized seafloor methane hydrate, constituting a potential trigger for the large methane pulse that is argued to have contributed to the PETM warming. It might have generated other ocean changes as well. Although the extinction event seems to have been confined to the deep ocean, the PETM did cause some notable changes in the upper ocean. There were widespread blooms of dinoflagellates in coastal ocean regions. An ancient form of red tide algae, these blooms were likely favored by what is known as eutrophication: the increased continental runoff would have delivered increased nutrients such as nitrogen into the coastal regions, leading to large outbreaks of dinoflagellates. As with modern red tides, the blooms of dinoflagellates would soon run through their boom-and-bust cycle, dying, decomposing, and consuming ocean oxygen in the process, thereby threatening other sea life including fish populations. As today, warming ocean waters would have exacerbated these dangerous and deadly occurrences.[19]

The Silurian Hypothesis

Unlike the current warming, we know that the PETM episode of abrupt warming was natural in origin. Or do we? One of the studies discussed earlier suggested that there were likely two distinct pulses of carbon input into the system. The second was consistent with a massive input of methane released in response to warming. But the carbon source that triggered the initial warming? The study could not pin it down. So, let's have a bit of fun for a minute. And while we're doing that, we'll keep in mind the fact that science often gains insight by ruling out what cannot be true and what cannot have happened. That is the spirit in which we will embrace the Silurian hypothesis. I'll begin by telling a story from my childhood. Growing up as an American in the 1970s, I watched religiously a children's TV series called *Land of the Lost*. It was an instant hit when it premiered in the fall of 1974. As an eight-year-old boy obsessed with dinosaurs and time and space travel, I felt that the show was custom made for me. I was hooked.

The program featured a family who found themselves trapped in a bizarre subterranean land inhabited by dinosaurs, Ewok-like ape people called Pakuni, and malevolent lizard people called Sleestak. The Sleestak descended from a once peaceful and advanced race of reptilian bipedal humanoids (called Altrusians), but degenerated over time into the primitive, barbaric individuals who inhabited the ruins of their once great civilization.[20]

The storyline for the first season was written by science fiction writer David Gerrold who, among other things, wrote the famous

"The Trouble with Tribbles" episode for the original *Star Trek* series. Other famous science fiction authors, such as Ben Bova, Theodore Sturgeon, Norman Spinrad, and Larry Niven, wrote episodes for the series.

Gavin Schmidt is a contemporary of mine who is currently director of the NASA GISS climate modeling laboratory, having taken over from former director James Hansen some years ago. When Schmidt was growing up in the 1970s across the pond in the United Kingdom, he, too, was watching science fiction programming, the BBC TV series *Doctor Who* to be specific. One episode featured lizard people, awakened by nuclear testing after 400 million years of hibernation. These intelligent, bipedal reptilians ruled over the dinosaurs, but were forced to hibernate deep within Earth's crust to escape a global catastrophe. They are called the Silurians (since we're only having fun here, we'll overlook the fact that the Silurian period actually predated reptiles by a hundred million years and dinosaurs by nearly two hundred million years).

Early 1970s TV and film was full of tales of collapsed ancient lizard civilizations. Why? I have some thoughts. The early to mid-1970s was the apex of environmental dystopianism. It gave us films like *Silent Running* and *Logan's Run* premised on scenarios of environmentally driven societal collapse. The 1973 film *Soylent Green*, starring Charlton Heston, was arguably ahead of its time. Premiering decades before widespread awareness of the climate crisis, it was premised on the devastating societal consequences of global warming. The story takes place, coincidentally, in the year 2022.

Then there's the 1968 dystopian film *Planet of the Apes*, which once again starred Charlton Heston, in the role of an astronaut who has found himself stranded on a planet ruled by intelligent, ape-like hominids. Toward the end of the film, he realizes that he had time traveled when he happens upon the archeological remains of his own civilization. It had destroyed itself through nuclear annihilation. The ape-like hominids had evolved to fill the void that was left behind.

Perhaps there is something archetypal about the notion of an intelligent civilization gone extinct under enigmatic circumstances. Maybe tales of this sort trigger something deep down in our own primitive lizard brains, some instinctual sense of our tenuousness on this pale blue dot we call home. Possibly such tales resonated with the dystopian environmental ethos of the 1970s, as we began to understand the threat posed to our planetary home by worsening air and water pollution, disappearing forests and habitats, and the nuclear-fueled Cold War that was simmering. Conceivably you're wondering what any of this has to do with the PETM.

What if an intelligent pre-human civilization like the Altrusians or Silurians existed tens of millions of years ago on Earth and extinguished themselves through catastrophic planetary warming, courtesy of an energy-greedy, fossil fuel–burning spree? Would we know it? This is the very thought experiment that was pursued by my friend and colleague Gavin Schmidt and his coauthor astrobiologist Adam Frank in a 2018 article titled, appropriately, "The Silurian Hypothesis."[21]

The project was a bit of an accident—as novel scientific pursuits, to be perfectly honest, often are. Adam Frank is a deeply inquisitive astrophysicist with a passion for addressing truly big questions, as I learned during a fascinating conversation over coffee on a chilly February day in 2019 while he was visiting the Happy Valley of State College, PA. He is also a leading advocate for the search for extraterrestrial intelligence (SETI), a continuation of the legacy of Carl Sagan, who cofounded the Planetary Society back in 1980 to advocate for the ongoing search for life in the cosmos.[22]

Back in 1961, at the very first scientific SETI meeting, the astrophysicist Frank Drake formulated a mathematical expression for the number of communicative civilizations in our galaxy as a product of various terms: the rate at which stars are produced, the number of planets per star, the fraction of those planets that would be habitable for life, the fraction of those on which life actually arises, the fraction of those that produce intelligent civilization, and the fraction of those that develop radio communication. The final factor is the typical lifetime of such civilizations. Carl Sagan was one of the ten scientists present at that meeting. He believed that the last factor was likely to be the limiting one. In other words, the key question, in Sagan's mind, was whether or not technological civilizations could avoid self-destruction. It is not unreasonable to speculate that these early musings on Sagan's part might well have prepared him for the later role he would play in the 1980s in the debate over the nuclear arms race.[23]

In 2017, Adam Frank paid a visit to Schmidt, a climate modeler. He was interested in the related astrobiological question of whether prospective industrial civilizations that arise on other planets might extinguish themselves through fossil fuel–driven warming. As I've learned from numerous conversations and collaborations over the years, Gavin Schmidt is an outside-the-box thinker. He's also a devil's advocate. So, he turned around and asked a stunned Frank a question of his own: "How do we know that a past civilization didn't already do this on Earth?" And so we come to what's called the Silurian hypothesis: how do we know that the rapid carbon spike behind the PETM warming, for example, wasn't the extinction-causing act of some ancient fossil fuel–hungry civilization? What sort of evidence might a subsequent civilization like us hope to find fifty or sixty million years later?[24]

The basic question has been asked numerous times before. The short novel *Nightfall*, written by Isaac Asimov in 1941, tells the story of an ancient civilization that was discovered through an analysis of the sedimentary record. The twist, though, is that it is actually the *same* species that is caught in a perpetual cycle of growth and collapse, driven by catastrophic wildfires. Only thin layers of ash in the sediments (and myths carried on through oral tradition) document the previous collapses. A later 1950 short story by Asimov, entitled "Day of the Hunters," tells the tale of an ancient, terrestrial, nonhuman civilization.

Science fiction author Larry Niven, who as you'll recall was a cowriter for *Land of the Lost*, wrote a story in 1979 called "The Green Marauder" that describes an intelligent species that lived on Earth hundreds of millions of years ago but apparently went extinct when Earth's atmosphere was "polluted" with oxygen (they were an anaerobic species). The story takes the form of a conversation in the Draco Tavern (also the title of the volume of short stories) where one of the interlocutors, ostensibly an alien from another planet, turns out to be a surviving member of this ancient species. Gavin Schmidt even wrote his own short story ("Under the Sun") premised on the Silurian hypothesis.[25]

Though scientists and novelists alike have speculated about such things for decades, Schmidt and Frank moved the ball down the field quite a bit by examining in detail the sort of geological and archeological evidence that would and would not be left behind by an intelligent, civilization-building species that drove itself to extinction through environmental destruction in the deep geological past. They note, for example, that the typical sorts of evidence we might imagine—continental-scale graves of human skeletons, collapsed edifices, cars and trucks, foundations of homes, etc.—simply wouldn't remain. Geological weathering and erosion along with plate tectonics would have destroyed any artificial structures and objects older than about ten million years.

There would be no direct evidence, in the sense of archeological sites or preserved artifacts, of an industrial civilization that only existed for a few centuries, a veritable fleeting geological moment. A very small fraction of living things ever become fossilized. So, if some race of reptiles or early mammals in the late Paleocene developed a civilization that lasted even for 100,000 years, let alone a few centuries, it would be easy to miss in the fossil record.

What evidence might we expect to find? We might see sharp coincident spikes in oxygen and carbon isotopes in preserved sediments, indicative of a rapid rise in greenhouse gases and temperatures. But that's precisely the sort of evidence we see with the PETM!

We might also expect to see a spike in nitrogen isotope ratios, indicative of the large-scale use of fertilizers, and increased anoxic zones in oceans, due to eutrophication, ocean acidification, and extinction of calcareous biota preserved in sediments. We could detect anomalous levels of lead, chromium, antimony, rhenium, and other mined metals in sediments. Interestingly, we do see these sorts of changes in the PETM and during other past episodes of rapid climate and environmental change due, for example, to increased erosion and continental runoff.

To be clear, Schmidt and Frank aren't actually suggesting that sentient lizards caused a warming spike fifty-six million years ago. Turns out, there is a perfectly good (and alas, far more mundane) explanation for what happened. Occam's razor, in the end, prevails. The authors concede that the hypothesis is almost certainly wrong. I asked Schmidt for the most compelling piece of counterevidence offered by his critics. His answer was: "Our experience with deep mining. These are metallic deposits that date back sometimes billions of years, and as far as I know, there is no evidence that they have been tapped previously." Yet the hypothesis isn't *obviously* wrong, either. It demands consideration and close examination.

Schmidt and Frank were merely posing the question of how future beings—including perhaps denizens of our planet millions of years hence—would know if a civilization like ours extinguished itself through environmental degradation and, specifically, a fossil fuel–driven abrupt warming event. The Silurian hypothesis was motivated by a deeper question that scientists like Adam Frank, David Grinspoon, Carl Sagan, and even the great physicist Enrico Fermi have long pondered: Is there other life out there? If so, why haven't we heard from it? Some have speculated that intelligent civilizations, perhaps, tend to sow the seeds of their own destruction through environmental ruination and warfare. And it's certainly worth asking: Is that our inclination? And if so, can we defy that impulse?

Hothouse Feedbacks

If we can rule out fossil fuel–hungry bipedal reptiles as the cause of the PETM, then whom should we blame? How about Iceland? It is, after all, a hot spot. Confused? Let me explain.

The island of Iceland formed around sixty million years ago from a plume of hot molten magma that rose up through the ocean from a "hot spot" in Earth's mantle, the basaltic lava eventually cooling into rocky crust, forming an island in the center of the North Atlantic just below the Arctic circle. Iceland remains perched above the mantle plume and hot spot, but it is also located along the Mid-Atlantic Ridge, a seafloor-spreading center where two tectonic plates—the Eurasian and North American plates—are diverging at roughly the rate your fingernails grow (about an inch per year). The net result is lots of geothermal heat and lots of volcanism.[26]

I have witnessed and experienced it in my visits to Iceland over the years. I've seen not just the geothermally heated geysers, but Geysir itself—the great geyser in southwestern Iceland that gave rise to the term in the first place. I've viewed the magnificent volcanoes including Laki, whose 1783 eruption famously led to dry fogs and

Figure 14. My visit to the Mid-Atlantic Ridge in Iceland (October 2013).

frigid conditions in Europe that summer, and Eyjafjallajökull, one of the most active volcanoes in the world and the *most* active volcano whose name cannot be pronounced. I swam in the so-called Blue Lagoon, an artificial pool made from the waste water of a geothermal power plant with purported healing qualities. And I've visited Thingvellir National Park in southwestern Iceland, standing atop the Mid-Atlantic Ridge, looking down into the canyon below formed by the two spreading plates.

The North Atlantic Igneous Province, or NAIP, is a large accumulation of igneous rocks—and an extremely rich reservoir of carbon—that is located above the magma plume underlying Iceland. It appears to have been formed during two massive episodes of volcanism between sixty-two and fifty-five million years ago, coincident with the initiation of seafloor spreading and opening of the northeast Atlantic Ocean. Recent studies have shown that an extended episode of volcanic outgassing tapping into this huge carbon reservoir might well have fueled the PETM.[27]

A 2017 study in *Nature* argued that a whopping 10,000–12,000 GtC of atmospheric carbon released from ongoing eruptions in this region over a 50,000-year timeframe was the primary driver of PETM warming. The authors used an innovative approach known as data assimilation. (I've used this innovative tool in my own paleoclimate research; it involves merging real-world data with a numerical model to come up with the likeliest scenario.)[28]

The authors estimated that there was most likely about 11,200 GtC released and that ninety percent of the carbon came from slow volcanic outgassing at a rate of about 0.6 GtC per year. That's more than a factor of ten smaller than the current rate of input from fossil fuel burning, confirming that though the PETM carbon release might have been "rapid" from a geological standpoint, it was slow compared to the generation of carbon pollution by humans today.

The *Nature* study authors estimated that the remaining ten percent of the carbon released during the PETM could have come from methane hydrates. So, hothouse, methane-related, positive feedbacks could have added modestly to the PETM warming, but they did not play a primary role.[29]

Could there have been any other hothouse feedbacks that kicked in as Earth approached the downright infernal average PETM temperature of 90°F? Perhaps. With apologies to Joni Mitchell, let's look at clouds from both sides now. And by both sides, I of course mean CO_2 levels both above and below 1200 ppm. A 2019 study in the journal *Nature Geoscience* led by Caltech researcher Tapio Schneider suggested that something very unusual happens to cloud behavior when we cross that particular CO_2 threshold.[30]

Schneider and colleagues focused on the role of stratocumulus decks, vast layers of low, reflective clouds that are found over the tropical and subtropical oceans, covering twenty percent of the area between the latitudes of 30°S and 30°N (which is half the surface area of the planet). The regions include large swaths of the eastern Pacific Ocean off the California, Mexican, Peruvian, and Chilean coasts and large swaths of the eastern Atlantic Ocean off the north and south African coasts. Using a high-resolution model of the atmosphere that, unlike standard climate models, can resolve individual clouds at scales as small as 150 feet, they found, quite remarkably, that the cloud layers essentially evaporated as the climate warmed beyond the threshold reached at CO_2 levels of 1200 ppm. The disappearance of clouds and dramatic increase in the absorbed sunlight led to an enormous temperature jump of 14°F.

The stratocumulus clouds become thinner with greenhouse warming, leading to an amplifying feedback with less reflection, more surface warming, and further thinning of the clouds. Eventually, a tipping point is reached where the clouds disappear altogether. It's much like the ice albedo feedback we encountered in Chapter 2, but with clouds, rather than ice, playing the key role. There is also hysteresis behavior of the sort we've encountered before with Daisyworld. The clouds dissipate in their model when the CO_2 level exceeds 1200 ppm, but once they're gone, they only reappear when CO_2 levels drop substantially below that level.

This destabilizing feedback may have contributed to hothouse climate behavior in the past. An obvious candidate is the PETM when CO_2 levels might have reached as high as 2500 ppm. The authors warn that this positive cloud feedback mechanism could appear as an unwelcome surprise as we continue to raise CO_2 levels through fossil fuel burning.

It is worth noting that state-of-the-art climate models show no hint of this sort of threshold-like cloud behavior, even at CO_2 levels as high as 9000 ppm. Schneider and his team argue that this is due to shortcomings in the way that the climate models parameterize cloud behavior. Other researchers are unconvinced. They have criticized the highly idealized nature of the experiments. Chris Bretherton of the University of Washington, for example, argues that cloud behavior transitions "happen at different times in different concentrations of CO_2 in different places. That would smooth it all out." In other words, there wouldn't be any single tipping point or abrupt, step-like disappearance of clouds and subsequent spike in global temperatures. Instead, there would likely be a more modest, and more gradual, change. Though that criticism is fair, it doesn't actually contradict Schneider's basic hypothesis. It just modifies it. Where the effect might be most apparent in the real world is in a higher level of climate sensitivity for very warm greenhouse climates like the PETM. And there is indeed some evidence for this, as we shall now see.[31]

Climate Sensitivity Revisited

Climate sensitivity, as you may recall, is the amount of warming eventually reached after we double the concentration of greenhouse gases. In the following chapters we will look at evidence from the more recent past, such as the Last Glacial Maximum (LGM) or the Common Era of the past two millennia, for insights into climate sensitivity. These examinations are useful, in part, because we have more accurate estimates of both temperatures and climate drivers during these more recent times. But the conclusions we draw from these periods are laden in caveats. Climate sensitivity is not a universal quantity. It involves feedback processes that are in general not the same for cold and warm global climates. Cold global climates, for example, are more likely to be impacted by ice cover–related albedo changes, whereas warm global climates are more likely to be impacted by carbon cycle feedbacks—including permafrost melt and methane release—and, as we've seen, possible warm-climate cloud changes.

This sort of asymmetry could mean that climate sensitivity estimates derived from past and recent *cold* climates are not especially instructive when it comes to future potential greenhouse warming. To find an analog for a hothouse climate with CO_2 levels of 1200 ppm or higher, which we could reach by the end of the century in a worst-case emissions scenario, we must go back to at least the early Eocene, around fifty million years ago, and perhaps back to our best hothouse analog, the PETM.

So, let's re-examine the PETM and see what it might tell us about hothouse climate sensitivity. Just prior to the PETM, during the late stages of the Paleocene, atmospheric CO_2 levels were roughly 870 ppm. At the height of the PETM, they were closer to about 2200 ppm. That's an increase by a factor of around 2.5, or one and a third doublings. Because the warming was in the range of 9–11°F, we can calculate an equilibrium climate sensitivity (ECS) of 6.7–8.1°F. You might rightly ask: aren't we measuring Earth system sensitivity

(ESS), which accounts for the role of slower changes such as ice sheets? To the extent that the main temperature increase was relatively rapid, taking place over thousands rather than hundreds of thousands of years, the slowest components of the Earth system weren't in play. And as there was no permanent ice on the planet, changing ice sheet extent was not a factor then. Comparing the PETM ECS estimate to the most likely current day value of 5°F seems to indicate greater climate sensitivity during this hothouse period.[32]

Recent studies provide additional support for that. A 2016 study, for example, found that ECS increased from 8.1°F prior to the PETM (when global average temperatures were about 79°F) to 9°F during the PETM (temperatures were about 90°F). Comparing to estimates from the deep cold of the LGM and from the modern era, the authors argued for a systematic increase in ECS with overall warming. A more recent study came up with slightly different numbers, but with the hothouse climates still exhibiting substantially higher sensitivities than the cooler climates. Other recent studies confirm the tendency for ECS to increase with warming, though the details continue to be debated in the technical literature.[33]

Lessons Learned

What does the PETM tell us about today's climate crisis? Though the main carbon spike that led to the PETM happened over less than 20,000 years, CO_2 concentrations and global temperatures remained elevated at hothouse levels for at least 100,000 years, a factor of five times longer. That underscores the tendency for carbon dioxide to remain in the atmosphere for a very long time once it's put there. And that same factor of five applies today: the carbon pollution we've already put in the atmosphere will keep CO_2 levels and temperatures elevated for more than a thousand years, absent substantial removal of atmospheric carbon. That means that our fragile moment could ultimately be threatened *even if* we stop burning fossil fuels now and rapidly bring our carbon emission to zero.

So, we might have to turn to methods—both natural (like massive reforestation and afforestation) and artificial (like prodigious carbon capture and sequestration technology)—that can help cool the planet back down. The technology for the latter is at best at the proof-of-concept stage at present. It is not viable at scale, but decades down the road it may become an important tool. Yet, as I emphasized in *The New Climate War*, promises of carbon capture and sequestration, and other proposed techno-fixes, must not be used as an excuse by polluters for delay and continued business as usual. Decarbonizing our civilization now remains task number one, and it can be accomplished with existing renewable energy and storage technology, efficiency measures, and smart grid technology. We currently lack only the political will, not the know-how, to accomplish this task.[34]

How about the risk of "methane bombs" so often cited by those who insist that runaway warming is now inevitable and climate action essentially futile? The PETM argues against such a scenario. Though it is an imperfect analog for modern warming, to be certain, what played out in the PETM seems to allay concerns about runaway methane feedbacks. Baseline temperatures prior to the PETM were more than 18°F warmer than today, and an apparent shift in the ocean conveyor favored burial of warm water (a collapse of the conveyor today wouldn't have that same effect). The deep oceans were consequently considerably closer to the temperature threshold at which methane hydrate becomes destabilized. Yet, there was no catastrophic release of methane hydrates. Despite ongoing accounts even in the mainstream media that imply otherwise, there was no PETM "methane bomb." The methane hydrate feedback during the PETM appears to have been at most ten percent of the total carbon release.[35]

There are caveats, of course. The *rate* of warming today is more than ten times greater than the PETM warming, and there is evidence that the destabilization of methane hydrates might be greater in a scenario of more-rapid warming. Much of the methane potentially in play today is located along the circum-Arctic continental shelves in the form of subsea permafrost. This additional methane source didn't exist in the PETM because Arctic sea ice didn't appear until roughly forty-seven million years ago. Subsea permafrost thawing and dissociation of methane hydrate have been fostered by warming and coastal inundation due to rising sea levels since the end of the last ice age. It is possible that more could be liberated due to continued human-caused warming. There is no evidence, however, that this is happening currently.[36]

That doesn't mean that methane isn't a problem today. It is. But not as a climate feedback. Rather as a human-caused climate driver. Methane is both a fossil fuel (in the form of natural gas, which produces CO_2 when burned) and a greenhouse gas. As noted earlier, we are witnessing a rise in methane concentrations due to natural gas extraction, livestock, and farming. The methane emissions appear to be from *us*, not some feedback cycle. Given that the rise in methane is responsible for about twenty-five percent of the warming in recent decades, reducing human methane emissions must be part of any comprehensive plan for addressing the climate crisis.[37]

Can modern-day humans warm the planet enough through fossil fuel burning to trigger other hothouse climate feedbacks? Perhaps, like the PETM, to a point where it's literally too hot for human beings? Such a scenario seems unlikely though not impossible. Earth's temperature today is about 60°F. Current policies alone, even without additional climate action, are projected to warm the planet at most about 5°F. That still places global temperatures below 65°F. That's 25°F cooler than the PETM and far below levels where feedbacks related to speculated tipping point changes in cloud cover are likely to kick in. But that doesn't mean that it won't be catastrophic. That's warm enough that parts of the planet could become unlivable. And we are already seeing a substantial increase in climate-related fatalities. In 2003, a record heat wave in Europe killed 30,000 people. The event was made at least ten times more likely by human-caused warming. The Pacific Northwest "Heat Dome" in the United States and Canada in June 2021 led to more than a thousand fatalities. As my colleague Susan Joy Hassol and I put it in a *New York Times* op-ed: "It [has been] called a once-in-a-millennium event, which means you might have expected to witness it once during your lifetime—if you happen to be Methuselah of biblical fame." We entitled our commentary "That Heat Dome? Yeah, It's Climate Change."[38]

It's also possible that current model projections underestimate the potential warming, due, for example, to an imperfect representation of carbon cycle feedbacks. Methane feedbacks, as we've seen, seem unlikely to be a major player, at least given climate policies already in place. But I worry about other wild cards such as drought-fueled CO_2-generating wildfires, like those we've witnessed in recent summers from the United States to Australia, and other potential carbon cycle feedbacks that might not be fully captured in current-generation climate models. Some climate scientists argue, for these and other reasons, that we can't yet rule out worst-case scenario CO_2 levels of about 1200 ppm by the end of the century even with current policies. Other experts dispute that claim, arguing that the most credible policy projections limit CO_2 levels to about half that much. I tend to side with that latter view, but in any case, that's still too much.[39]

The fact that the rate of warming today far exceeds that during the PETM presents its own unique challenges. We saw that mammals and other species migrated away from regions that became too warm, or literally—as in the curious case of the shrinking horses—adapted to the warmth. But the rate of warming today, as we've seen, is more than ten times greater, exceeding the rate at which plants and animals can be expected to migrate or adapt. And adding insult to injury, we've built all sorts of obstacles—in the form of cities, highways, and other infrastructure—that stand in the way of likely migration routes.

Now what about a *worst-case* scenario, where we actually regress, reversing the climate policies we've already enacted, proceeding instead to burn all of the reasonably accessible fossil fuel reserves? The state-of-the-art model projections used in the most recent assessments of the Intergovernmental Panel on Climate Change (IPCC) indicate a most-likely warming, in that case, of about 7°F by 2100, plateauing in 2300 to an approximate 14°F warming. That is a huge, devastating amount of warming, but it's not a "runaway" greenhouse scenario, nor is it a PETM hothouse scenario. What if we instead take the *most extreme* end of the IPCC simulation range? In that worst-case scenario we're looking instead at as much as 11°F by 2100, plateauing to about 23°F in 2300. That would put global average temperatures at around 83°F two centuries from now. That scenario is extremely unlikely as it assumes a reversal of climate policy progress already made and is based on the most extreme of the more than fifty climate models analyzed by the IPCC. Though it still falls several degrees short of the PETM, it's uncomfortably close, and possibly hot enough that much of the planet would be uninhabitably hot for humans and other large mammals. So yeah, if we try really, really hard, we could make at least most of this planet unlivable for human beings.[40]

My friend and colleague Matt Huber, a leading climate scientist at Purdue University who has published seminal work on both the PETM and contemporary heat stress, puts it this way: "Lizards will be fine, birds will be fine." But us? Not so much. Huber and his coauthor Steve Sherwood of the University of New South Wales (UNSW) in Australia note that in such a scenario we'd experience severe challenges from heat stress and other climate impacts, combined with inadequate infrastructure, including failing power grids that might not protect us from deadly heat. They note that "the power requirements of air conditioning would soar; it would surely remain unaffordable for billions in the third world and for protection of most livestock . . . it would regularly imprison people in their homes; and power failures would become life-threatening."[41]

Consider this account by the *New York Times* of an actual blackout in New York City in July 1977: "On a steamy July night in 1977 . . . New York City plunged into darkness . . . left without power for 25 hours. It became a defining event. Looting and arson spread through the streets, resulting in 3,800 arrests and millions of dollars' worth of damage." Such events could become commonplace in the major cities of the world. It is not difficult to imagine a scenario of societal collapse. If this all sounds a bit like the aforementioned *Soylent Green*, it's because our world could indeed in a worst-case scenario be a bit like the film *Soylent Green*.[42]

The good news? Even in a scenario where we simply maintain climate policies already on the books and do little if anything beyond that, it is extremely unlikely to give us anything remotely resembling a PETM-like dystopia. The bad news? In the absence of a ratcheting-up of governmental climate mitigation efforts, we could—if the upper-end range of the model projections proves correct—see as much as 7°F of warming by the end of the century: by any measure, a world of hurt. If the PETM isn't a great analog for what that might look like, what is? Perhaps some ancient but more recent climates that weren't quite as hot and where CO_2 levels were comparable to those we may reach in such a climate policy scenario. That's the topic of our next chapter.

A Message in the Ice

Civilization is like a thin layer of ice upon a deep ocean of chaos and darkness.

—WERNER HERZOG, *Herzog on Herzog*

The dinosaurs saw the end of their own moment sixty-six million years ago with the Chicxulub asteroid impact. The geological era that followed is known as the Cenozoic. It's sometimes referred to as the Age of Mammals. Without those pesky dinosaurs around, we eventually got to run the show. The PETM and its carbon-driven warming spike occurred early in this era. But it was followed by a broader interval of elevated warmth in the early Eocene (from fifty-five to forty-six million years ago) sometimes called the Eocene Climate Optimum. Then Earth began to cool down.

The cooling trend, as we learned back in Chapter 1, was a consequence of a gradual drop in atmospheric carbon dioxide levels over millions of years, due to plate tectonics and, especially, the collision of the Indian subcontinent with Eurasia. That collision created the dramatic relief of the Himalayas, a strengthening of the Asian monsoons, increased rainfall, more weathering of silicate rocks, and a steady drawdown of atmospheric carbon dioxide. The greatest beneficiary was the ice, which had a chance to re-emerge after a very long hiatus. The planet slowly began to better resemble the one we live on today, providing critical possible analogs for the climate we might be heading toward in the near future. Among the key lessons to be learned is just how much loss of ice, and how much resulting sea level rise, we might be in for.

The Ice Age Cometh

As you may recall from Chapter 1, the warm, humid greenhouse climates of the early and mid-Eocene produced arctic alligators, polar palm trees, and the humid forested landscapes that allowed our arboreal early primate ancestors to thrive across the Northern Hemisphere. The climate remained 9–27°F warmer than today, and there was no ice at either pole. The global sea level was about 260 feet higher than it is today. Now, the continental and ocean basin geography was a bit different from what it is today, so the translation isn't perfect. But that's roughly as much sea level rise as we would see if we melted the two remaining continental-scale glaciers—the Greenland and Antarctic Ice Sheets—a scenario that is possible depending on the fossil fuel emissions pathway we follow.[1]

By the latter stages of the Eocene, roughly forty-five million years ago, ice finally returned to our planet having been absent for more than 200 million years. It first formed in the coldest regions of Antarctica. Eleven million years later it rapidly spread across much of the Antarctic continent. Earth had undergone a rather abrupt transition from the greenhouse world of the Eocene to the icehouse world of the Oligocene. The so-called Eocene-Oligocene transition of thirty-four million years ago saw a global temperature decrease of about 8°F in the space of about 400,000 years. Sea levels dropped and the continents got drier. If the PETM was a rapid natural *warming* event, here was a rapid natural *cooling* event.[2]

As with other abrupt climate shifts, there were widespread extinctions. A broad swath of life, from plants to marine invertebrates like forams, to gastropods and bivalves, and to mammals such as rodents, saw species disappearance in excess of fifty percent. But, as we are almost tired of saying by now, *there are always winners and losers.* A diverse group of hoofed mammals known as ungulates was decimated. But the larger species among them, like horses, rhinos, cattle, pigs, giraffes, camels, sheep, and deer, who were able to fully exploit

Figure 15. Artist depiction of a scene from the Oligocene with *Meso-hippus bairdi*, a browsing, short-necked, three-toed horse (order Perissodactyla, family *Equidae*).

the rapidly expanding grasslands and the replacement of dense forests with more open woodland thrived and survive to this day. Many of the extinctions involved animal groups that had lived in warm, high-latitude regions of Eurasia that cooled off rapidly. Yet, there were widespread extinctions even in tropical and subtropical African environments, including various species of lemurs and monkeys. Our own ancestors—species belonging to the suborder anthropoid—obviously made it through. Indeed, they presumably benefited from their growing intelligence and behavioral plasticity and the lucky break of less competition from extinct competitors.[3]

What drove the dramatic Eocene-Oligocene climate shift? As we've seen with other similar episodes, there were both a driver—tectonically induced changes in CO_2 concentrations—and amplification by feedback mechanisms. The collision of the Indian Plate with the Eurasian Plate close to the Eocene-Oligocene boundary led to the initial uplift of the Himalayas. The increased relief and forced rising of air it induced helped, as we've already learned, establish the South Asian summer monsoon, leading to increased continental precipitation, enhanced silicate weathering, and a downturn in CO_2 over millions of years. That alone, however, cannot explain the abrupt, step-like cooling that occurred in just a few hundred thousand years. There are several, nonlinear climate feedback mechanisms—relevant today—that likely came into play, helping trigger the rapid transition toward glaciation and cooling.

There are, of course, the basic feedbacks associated with cooling, the most fundamental of which is the ice albedo feedback we talked about in Chapter 2. Proxy data from preserved carbonates during that time indicate that CO_2 had probably dropped to about 750 ppm by the beginning of the Oligocene. That's about 2.7 times the preindustrial level of 280 ppm. That factor of 2.7 appears to have been the tipping point for glaciation of the Antarctic continent.[4]

A 2003 study in *Nature* by a pair of leading paleoclimate modelers (and former colleagues), Robert DeConto of UMass Amherst and David Pollard of Penn State, used a detailed model of Antarctic ice sheet behavior combined with a global climate model to investigate this transition into glaciation. They showed that an ice sheet could form rapidly over most of Antarctica once CO_2 levels drop to roughly the levels encountered at the Eocene-Oligocene boundary. They found that an additional *ice elevation* feedback was important in explaining the abrupt, tipping point–like growth of the ice sheet. Here's how it works: As an ice sheet grows in size, it also grows in height. Because atmospheric temperatures decrease with altitude, the top of the ice sheet is colder than the bottom, and precipitation that might have reached bare ground as rain instead lands as snow on top of the ice sheet, further increasing the accumulation of snow and ice.[5]

In these model simulations, large-scale glaciation occurred when CO_2 levels dropped to about three times preindustrial levels. That's a bit higher than the level (2.7) suggested from paleoclimate data, indicating that glaciation appears to have taken place later than their model predicted. What might be responsible for the discrepancy? Well, at roughly the same time India was colliding with southern Asia in the Northern Hemisphere, the Drake Passage—the geographic land gap between Cape Horn of South America and the northern tip of Antarctica—was just beginning to open, connecting the south Atlantic and south Pacific Oceans. The main surface current in the southern ocean could then simply follow the band of westerly winds, encircling Antarctica rather than colliding with it. That reduced oceanic warming of the continent by perhaps twenty percent, enhancing the cooling and glaciation. When the researchers included that effect in the calculation, the CO_2 factor dropped from 3.0 to 2.5. In other words, the two simulations nicely bracket the proxy observations, suggesting that the model is at least more or less capturing what happened in the real world.

There could nonetheless be other feedback mechanisms at work that are still not accounted for in these studies' modeling framework, mechanisms that might add to climate sensitivity today. Let's look back to the early Eocene, around fifty million years ago, at the beginning of the Eocene-Oligocene transition. The Arctic was frost-free and CO_2 levels are estimated to have been very high, around 2000 ppm. Current climate models only show a frost-free Arctic at much higher levels of around 4000 ppm. One possibility is that the hothouse cloud effect identified by Tapio Schneider and colleagues, as discussed in the last chapter (which involves the evaporation of reflective stratocumulus cloud decks at sufficiently high CO_2 levels), isn't accounted for in these existing climate models. This could explain why the current models underestimate the warmth of hothouse climates. It might also help explain the *abruptness* of climate transitions such as the Eocene-Oligocene glacial transition thirty-four million years ago.[6]

To understand how, let's return for a moment to the nonlinear phenomenon of hysteresis—the defective shower knob effect we discussed in Chapter 2. We have seen that hysteresis is present in various aspects of our climate, including our biosphere. Recall the charred remains of a relict loblolly pine forest that I encountered during a drive from Austin to College Station, Texas, in September 2012, a year after they'd been destroyed by record 2011 wildfires. This is just one example of numerous relict habitats around the world, forests and ecosystems that are out of equilibrium with today's climate. They exist now because they're holding on, if tenuously, despite the challenges of our changing climate. But if you destroy them, they don't return—other forest ecosystems more competitive in today's climate take hold in their place, even with climate conditions that are the same as they were prior to the disturbance.

Hysteresis can also help us understand the asymmetrical way the climate system behaves when it is in the warming phase versus the cooling phase. In their admittedly idealistic simulations, Tapio Schneider and colleagues found that the cloud effect kicks in at about 1200 ppm when the CO_2 is increasing, but only disappears at a considerably lower level—perhaps as low as 750 ppm—when the CO_2 is on the downturn, as it was at the Eocene-Oligocene boundary. So, it's possible that this effect could have contributed to the rapid cooling that took place when CO_2 levels dropped to about 750 ppm at the Eocene-Oligocene transition.

Hysteresis is also present, as we have seen, in other basic ice feedback mechanisms. In Chapter 2, we used a relatively simple type of climate model—the Budyko-Sellers model—to investigate the transition of Earth's climate both into and out of ice ages. We used the model to gain insights into the phenomenon of Snowball Earth, but its applicability is far broader. And it's relevant here in the trend toward glaciation. The model, you'll recall, allowed us to vary the heating (which can come from the Sun or CO_2 levels), demonstrating that the transition from an ice-free to icy planet happens at a lower level of heating than the reverse transition from an icy to ice-free planet. Hysteresis.

One way to think about this fact is that the ice albedo feedback has a much bigger effect overall on a planet with a lot of ice. It's more difficult to melt ice when there's a lot of it around. An icy Earth wants to stay icy, and an ice-free Earth wants to remain that way. Gaia and Medea once again battling it out.

In a 2005 study, Pollard and DeConto were able to quantify this glaciation effect for Antarctica. They included in their model simulations the shorter-term Earth orbital cycles that we discussed in Chapter 1. These cycles operate on timescales of tens of thousands of years and impact the distribution of solar heating over Earth's surface, including the critical polar regions where ice sheet growth is initiated. It's important to include these short-term glacial/interglacial cycles on top of the longer-term driver of dropping CO_2 levels, because ice growth is threshold-dependent. A "glacial" Earth orbital configuration that favors less solar heating in the summer at high latitudes can add to the cooling effect of decreasing CO_2 levels, pushing the climate past the critical threshold for polar ice growth. What they found was that including the effect yields a more accurate prediction of glacial inception, with the CO_2 level for initiating the Antarctic Ice Sheet about fifteen percent lower than the CO_2 level for melting it.[7]

If the entire Antarctic Ice Sheet were to melt, it would raise global sea levels by about 200 feet (60 meters). All of the coastal cities, and forty percent of the world population, would be inundated. The good news, as we saw earlier, is that the CO_2 concentration required to melt the Antarctic Ice Sheet looks to be higher than the estimated 750 ppm level required for it to form thirty-four million years ago. The bad news is that it's not much higher, probably around 850 ppm. That's roughly the level that existed during the early Paleocene, just prior to the PETM—a level that we will easily reach if we choose to extract and burn all known fossil fuel reserves.

The Pliocene Omen

The CO_2 drawdown and cooling continued on through the Oligocene into the subsequent Miocene, which began twenty-four million years ago. The cooler, drier conditions led to the further retreat of subtropical forests and rise of less water-hungry woodlands, which favored ape species such as orangutans, gorillas, chimps, and our own closely related ape ancestors. Woodlands gave way to grasslands as CO_2 levels dropped further, and the cooling and drying continued on. The grasslands expanded further, creating a niche for the big-brained, bipedal hunter-gatherers of the vast African savannas. Just a bit more than five million years ago, we entered the Pliocene. Or if you like, "the Age of the Hominins"—proto-humans.

A variety of evidence from carbon isotopes to leaf stomata—the pores found in ancient leaves—suggests that Pliocene CO_2 concentrations were similar to, though probably slightly lower than, today, somewhere between 380 and 420 ppm (they are at about 420 ppm today). That might seem to make the Pliocene a pretty good analog for today. And yet it's not. The planet was warmer then, likely 1.5–3.5°F warmer globally than today. But temperatures were as much as 18–36°F higher than today at polar latitudes, while tropical surface temperatures were generally similar to current day.[8]

The combination of substantial warming at high latitudes and little warming at lower latitudes is not well captured by climate model simulations. It is in fact symptomatic of a larger, long-standing "equable climate" problem observed for other past warm periods like the Eocene epoch and the Cretaceous period. There have been various hypotheses regarding possible modes of heat transport to higher latitudes by the ocean or atmosphere that might be underestimated by, or simply missing in, climate models. One intriguing possibility was suggested some years ago by MIT hurricane researcher Kerry Emanuel and colleagues. They hypothesized that a warmer climate might support greater tropical cyclone activity, which would enhance the wind-induced mixing of the upper ocean, increasing the efficiency with which the ocean transports heat from low to high latitudes. In any case, this constitutes an important caveat. To the extent that the models don't do a good job capturing this behavior in the past, they might not be capturing similar features of future warming, including the amplified high-latitude warming so relevant to ice sheet collapse and coastal inundation.[9]

With such warm polar regions, ice sheet coverage during the Pliocene was substantially reduced compared to today. We learned in Chapter 1 that the Antarctic Ice Sheet (AIS) formed during the Oligocene. But it's useful to draw some distinctions here between the high-elevation portion of the ice sheet—the East Antarctic Ice Sheet (EAIS)—and the low-elevation portion—known as the West Antarctic Ice Sheet (WAIS). The two distinct ice sheets are separated by the Transantarctic Mountains, falling mostly in the Eastern and Western Hemispheres, respectively, hence their names (though "east" and "west" lose their meaning when you approach the pole).

Ice core evidence suggests that the WAIS formed later than the EAIS, during the mid- to late Miocene, first as a terrestrial ice sheet. As the underlying bedrock began to drop due to a combination of plate tectonics and glacial erosion, the ice sheet transitioned during the Pliocene to a more dynamic, marine-based ice sheet—an ice sheet that rests on land lying below sea level, making it susceptible to melting from an influx of seawater. The ice retreated during warm interglacial periods and expanded during glacial periods, which has implications for the fragility of this ice today. By the very late Pliocene or early Pleistocene, two to three million years ago, the Greenland Ice Sheet (GIS) formed and the intermittent advances and retreats of the WAIS increased in frequency.[10]

Of the roughly 200 feet of sea level tied up in the Antarctic Ice Sheet, about 180 feet is stored today in the much larger EAIS. Only about twenty feet is tied up in the WAIS. And another twenty feet is tied up in the GIS. So, you might be forgiven for thinking that it's the EAIS about which we should be most concerned. But in reality, it is the WAIS that poses a much greater sea level rise threat—at least in the foreseeable future.

We are unlikely to breach the approximate 850 ppm CO_2 level that appears necessary to melt the EAIS *if* we take meaningful steps to reduce carbon emissions. It is relatively stable, as most of it rests on bedrock that lies well above current sea levels. The WAIS, on the other hand, is much less stable. That has to do with what's known as the grounding line—the boundary where the portion of the glacier that is above water is balanced by the portion that is underwater. Beyond the grounding line lies the ice shelf, floating ice from which icebergs can calve into the ocean. The WAIS mostly sits on bedrock that is below sea level, which means that ocean water can destabilize the ice by eroding it and melting it back to the grounding line. As the grounding line continues to retreat, more ice shelf is exposed to the water, leading to the warming, thinning, and fracturing of the ice shelf, which further destabilizes the inland glaciers through a loss of buttressing support. This feedback process, known as the Marine Ice Sheet Instability (MISI), can lead to rapid ice sheet collapse.

Sea level has been estimated to have been as much as 33–132 feet higher in the mid-Pliocene during highstands when Earth orbital cycles superimposed on the longer-term cooling trend favored peak polar warmth. The mid-range value is eighty-two feet. If you're paying close attention to the numbers, then you may have noticed a discrepancy. If we add the GIS (twenty feet) and WAIS (twenty feet), we get forty feet worth of sea level. That leaves a whopping forty-two feet of mid-Pliocene sea level unexplained—a pretty big deal given that today it would be the difference between a slightly flooded versus fully submerged Manhattan.

How do we resolve this discrepancy? A bit of it comes from mountain glaciers around the world, and the thermal expansion of oceans with warming. But that gives us at most three to six feet. Most of it is probably explained by something else—an error of sorts. It has to do with something known as *eustasy*. With ancient sea level reconstructions, scientists are seeking to estimate the so-called *eustatic* sea level rise—the rise in sea level due to changes in the amount of water in the ocean. But at any particular location, they are also measuring a contribution from what is known as glacial *isostatic* adjustment. Continents deform when there is a large ice load on them, like a continental ice sheet, and rebound slowly when the load is removed through melting, often taking thousands of years to do so. For example, during the Last Glacial Maximum roughly 21,000 years ago, there was a large ice sheet known as the Laurentide Ice Sheet over North America. If I were to have driven roughly one hour in the northeastern direction from my home in central Pennsylvania, I would have encountered its southern terminus.

After ice retreated and disappeared at the end of the Pleistocene roughly 12,000 years ago, the lithosphere—the Earth's crust and upper mantle—slowly rebounded, and is still doing so. It's known as post-glacial rebound. Parts of northern Ontario and Quebec that are proximal to the location of the greatest ice load, for example, are still rising about a millimeter a year. The New Jersey coast, on the other hand, is actually subsiding by more than a millimeter a year, due to the flexing of the lithosphere in the opposite direction. It has been estimated that the sea level discrepancy in mid-Pliocene sea level rise is due almost entirely to this effect, which has not been properly corrected for in sedimentary estimates of past sea level rise. Climate models tend to predict thirty or so feet of sea level rise relative to the present, consistent with the absence of the GIS and WAIS—or alternatively, the GIS, the marine portions of the WAIS (ten feet), and just a bit (three feet) from the least stable marine portions of the EAIS.[11]

The good news is that we can likely rule out an entirely submerged Manhattan. But sea level during the mid-Pliocene could well have been thirty feet higher than today, enough to displace a half billion people. Global temperatures were 3.6–5.4°F warmer, despite CO_2 levels that were probably slightly lower than those that prevail today. Does this mean that we are living on borrowed time? Is similar warming, ice sheet loss, and sea level rise already in the pipeline, and it's just a matter of time until we experience it? One still often encounters that argument. And yet it is probably wrong.[12]

To understand why, we need to revisit the concepts of equilibrium climate sensitivity (ECS) and Earth system sensitivity (ESS): the two different measures we encountered in Chapter 3 of how much warming we expect for a doubling of CO_2 concentrations. Today, CO_2 concentrations—at 420 ppm—are just under the halfway point between preindustrial (280 ppm) and twice preindustrial (560 ppm) levels. That means that the global temperature rise would most likely stabilize at roughly 2.7°F if CO_2 concentrations were held at current levels for several decades and the climate were to settle into a new equilibrium state.

However, that doesn't account for slow climate feedbacks, such as changing vegetation and ice sheet extent, that kick in if CO_2 is kept at this level for millennia, as it was during the Pliocene. That's what ESS is supposed to measure. University of Bristol paleoclimate modeler Daniel Lunt and colleagues argue for an ESS that is forty-three to forty-four percent larger than ECS based on comparisons of model simulations and paleoclimate reconstructions for the mid-Pliocene. That implies a warming of about 3.8°F if CO_2 levels are simply sustained at 420 ppm—basically, the levels that exist right now—for millennia. That's cause for concern given the devastating impacts that have been documented in a recent IPCC report on just 2°C (that's 3.6°F) warming. But that's still at the very lower end of the estimated range (3.6–5.4°F) of mid-Pliocene warming. Why was the mid-Pliocene so much warmer at similar CO_2 levels? Could there be a more fundamental problem with the way we are thinking about this comparison?[13]

The answer, once again, may be hysteresis. It is critical to understanding the behavior of the ice sheets in the past and, consequently, their likely behavior in the future. Estimating ESS from glacial climates of the late Cenozoic is problematic in the face of hysteresis, because there is no longer one unique sensitivity that describes the system at a given CO_2 level. As CO_2 dropped over the course of the Oligocene, Miocene, and Pliocene, we eventually reached a level similar to our current level of about 420 ppm. But due to unusual high-latitude warmth, there was no GIS or WAIS during much of the mid-Pliocene. (There's a bit of a chicken-and-egg problem here, as that high-latitude warmth was a result, at least in part, of the lesser extent of ice and decreased albedo.)[14]

The Lunt et al. study estimated ESS by modeling the mid-Pliocene, specifying the ice sheet and land surface distribution that existed at the time as well as the CO_2 level of about 400 ppm. They then measured the net change relative to a modern preindustrial baseline state (CO_2 of about 280 ppm). But here's the problem: that probably does not mimic what we would observe were we to simply keep CO_2 fixed at current levels and allow the climate to equilibrate over centuries or even millennia. The ESS estimated in this way for the mid-Pliocene is probably larger than the ESS that applies today, even though the CO_2 level is nearly the same. There's a good chance that the GIS would not disappear and that high northern latitudes would not warm up as much as they did during the mid-Pliocene if CO_2 levels are maintained at around 420 ppm.

A 2012 study in the journal *Nature Climate Change* led by paleoclimate modeler Alexander Robinson of the Potsdam Institute for Climate Impact Research estimated that the threshold summer temperature for melting the Greenland Ice Sheet is about 2.5°F higher than the threshold for forming it, which is, terrifyingly, only a bit higher than the warming today. A 2008 study suggests that the GIS formed during the late Pliocene, somewhere between 2.6 and 3 million years ago, when CO_2 levels had declined to somewhere between preindustrial and current levels. With this information, we can surmise that we have likely not yet crossed the threshold for GIS collapse, though we could be very close to it. Just a few tenths of a degree of warming could be enough to push us past that threshold. The WAIS is a bit more of a wild card. There is no clear evidence that it exhibits hysteresis, but there is evidence that here, too, we could be very close to the threshold for collapse.[15]

To confirm my own understanding of this critical matter, I turned to leading experts Robert DeConto and David Pollard. Pollard pointed out that the precise CO_2 range for hysteresis of the GIS is tricky to pin down given the sometimes-conflicting studies. But DeConto added that there has been no fundamental rethinking—the

GIS likely does, in fact, exhibit hysteresis. Lest one think this is merely a technical matter, it is actually the difference between whether or not New York City and heavily settled coastal regions around the world face inevitable, near-term inundation or not.[16]

The evidence seems to indicate that we're not yet past the point of no return for the collapse of the GIS. But we're probably getting perilously close, as we likely are with at least the marine portions of the WAIS. Together, that's enough ice loss to give us thirty or more feet of sea level rise, enough to displace hundreds of millions of people. How soon this might happen remains uncertain. But as we'll learn a bit later, recent research has led to a substantial upward revision in the likelihood of catastrophic sea level rise *this century*, absent substantial climate action.

The Pleistocene Glaciations

Carbon dioxide levels and global temperatures continued to drop as we transitioned from the Pliocene into the Pleistocene, a little more than 2.5 million years ago. Paleoceanographers think that the formation of the Isthmus of Panama about three million years ago, which blocked the flow of relatively fresh seawater from the Pacific Ocean to the Atlantic Ocean, led to the salinification of the Atlantic, a stronger ocean conveyor circulation, and a greater transport of snowfall-generating moisture to the high latitudes of the North Atlantic. The key factor, though, as we'll see, was the effect of continued drawdown of atmospheric CO_2, consequent cooling, and its impact on the behavior of increasingly large continental ice sheets.[17]

With an ice sheet already having taken hold in Greenland, ice began to spread southward into North America and northeastern Europe, giving us the Laurentide and Fennoscandian Ice Sheets, respectively. At its peak the Laurentide Ice Sheet (LIS) covered all of Canada and parts of North America as far south as southern Illinois and Indiana, New York City, and the northern half of my state of Pennsylvania. The Fennoscandian Ice Sheet (FIS) covered modern-day Scandinavia and neighboring regions of northern Europe.

These ice sheets waxed and waned over tens of thousands of years with the Earth's orbital cycles. The changes in ice volume are documented by oxygen isotopes from deep-sea sediment cores. We learned in earlier chapters that the ratio of stable oxygen isotopes (oxygen-16 to oxygen-18) in the preserved shells of calcareous biota varies with sea temperatures—being relatively depleted in oxygen-18 for warmer ocean temperatures. But we alluded to one complication: the ratio also depends on global ice mass. For hothouse climates, this is an insignificant factor, but for icehouse climates like the Pleistocene, it can be the dominant one. Yet, this is, as they say, a feature, not a bug, for it allows us to document the changes in global ice mass over time.

As we learned earlier, oxygen-18 is enriched in seawater (and thus calcite shells) when ice sheets form and grow. Because cold oceans and large ice sheets go together climatically, and each is characterized by enriched oxygen-18, a very clear and unambiguous signal of glacial cycles is captured by the oxygen isotope data.

The oxygen isotope data show a steady, long-term trend toward colder, icier conditions over the course of the Pleistocene. For the first million or so years, we see an oscillation between colder, icier (glacial) states and warmer, less icy (interglacial) states. The oscillations occurred with a roughly 40,000 year pacing, which we know is tied to the tilt angle of Earth relative to its orbital plane. As we learned in Chapter 1, the tilt angle today is about 23.5 degrees from vertical. However, the angle changes over time, between about 22 degrees and 24.5 degrees. On the low end, this means cooler summers and warmer winters at high latitudes—an ideal combination for the buildup of ice sheets, as warmer winters can increase the accumulation of snow while cooler summers reduce the amount of summer melt.

Something rather surprising happened between 1.25 and 0.75 million years ago, an interval that is known as the Mid-Pleistocene Transition, or MPT. As the climate grew colder and more ice-friendly, and the Laurentide and Fennoscandian Ice Sheets grew larger, the character of Earth's climate system fundamentally changed. We see a transition from fairly small-amplitude oscillations with a roughly 40,000 year periodicity and a smooth, sinusoidal shape to dramatic, sawtooth-shaped cycles with a roughly 100,000 year periodicity. The *glacial* intervals (the ice ages) were associated with very low carbon dioxide concentrations (about 180–190 ppm), extensive Northern

Hemisphere ice sheets, and cold conditions globally. The *interglacial* intervals were associated with higher carbon dioxide concentrations (about 280–300 ppm), minimal North American and European ice sheets, and relatively warm conditions. The peak-to-peak variation in global temperature was around 7°F.

A brief aside on how we can document past CO_2 changes so precisely. As we've seen before, they can be estimated from boron and carbon isotopes in carbonates from ocean sediments, but such measurements are very crude and uncertain. Because it is a well-mixed gas in the atmosphere, a precise estimate of CO_2 over time from one pristine location is representative of the global average levels of the gas. We have a long instrumental record today thanks to atmospheric chemist Charles Keeling who, at the behest of famed climate scientist Roger Revelle, began measuring CO_2 at the top of Mauna Loa in Hawaii back in 1958. We can get fairly precise estimates of past CO_2 changes much further back in time from gas bubbles trapped in ice from the cold, dry interior of East Antarctica. Vostok Station, located near the South Pole, is the coldest place on Earth—or, at least, the coldest place where we have reliable thermometer measurements—having recorded a temperature of –129°F on the mid-winter date of July 21, 1983. Because it is so cold, there's very little moisture in the air, and relatively little annual snowfall. Drilling down through the ice consequently yields an extremely long record in time.[18]

In 2012, a team of Russian scientists finally hit a freshwater lake at the bottom of the EAIS after having drilled a core 12,000 feet below the surface at Vostok, stretching 800,000 years back in time. The resulting Vostok ice core record documents not only CO_2 changes but also—through oxygen isotopes—a continuous record of temperatures in the interior of the Antarctic continent. That reveals regional temperature swings during the glacial/interglacial cycles that are more than twice the global average (about 18°F), demonstrating the familiar polar amplification of warming. The sawtooth-shaped temperature and CO_2 curves are in lockstep—which is not a coincidence.[19]

When Earth's orbit reaches peak eccentricity (ellipticity), as it did 125,000 years ago, the difference between the distance of Earth from the Sun during its closest approach over the course of the year and its farthest approach is amplified. That exaggerates seasonality, producing warmer summers that tend to melt whatever snow fell during winter, preventing any long-term buildup of ice over time. As Earth's orbital eccentricity decreases, like it did during the subsequent tens of thousands of years, the conditions for ice buildup become more favorable—CO_2 levels decrease, temperatures drop, and the planet slowly descends into an ice age. The last ice age peaked 21,000 years ago (known as the Last Glacial Maximum). The ice began to melt around 17,000 years ago once the eccentricity had again risen sufficiently, completing one roughly 100,000-year eccentricity cycle yielding, by 12,000 years ago, the current interglacial epoch known as the Holocene (the subject of the next chapter).

It is tempting, therefore, to simply blame the 100,000-year sawtooth cycles on eccentricity. But what then caused the *shift* during the MPT from small-amplitude, smooth 40,000-year glacial cycles that closely align with the variations in Earth's tilt, to asymmetric, around 100,000-year sawtooth-shaped oscillations aligned instead with eccentricity, characterized by slow cooling and ice growth followed by a rapid meltdown to warm interglacial conditions? This is a long-standing problem in paleoclimatology, with significant implications for our understanding of human-caused climate change. After all, the climate critics—who in general are way off base—would seem to actually have a point in this case: why should we trust the climate models currently used to predict future human-caused climate changes if they can't reproduce the most prominent signal in the paleoclimate record? It is for this reason that it really is worth spending a bit of time to review the historical evolution of our scientific understanding of the Pleistocene ice ages. And in the process, as a bonus, we'll learn some important lessons about how science works.

It starts with Serbian mathematician and astronomer Milutin Milankovitch in the 1920s. He argued that ice age cycles can be understood as the response to Earth orbital drivers such as precession (the "wobbling top" effect we discussed in Chapter 1, where one wobble takes about 23,000–26,000 years), obliquity (the tilt angle of that wobble, which varies with a periodicity of about 41,000 years), and eccentricity (with its about 100,000-year periodicity). Each influences the distribution of high-latitude summer insolation that determines whether snow and ice can persist through the summer and thus build up over time. The theory works well for describing glacial cycles on timescales of tens of thousands of years.[20]

There are challenges with the astronomical Milankovitch theory, however, when it comes to the 100,000-year ice age cycle. First and foremost, the eccentricity of Earth's orbit is very small. It remains *nearly* circular and the change in sunlight reaching Earth's surface is less than one percent. The changes in insolation from obliquity (Earth's tilt) during the key Arctic summer is more than ten times larger. Secondly, the ice ages just don't line up as neatly with the eccentricity data as they should. Finally, and most problematic of all, the standard astronomical theory doesn't explain the Mid-Pleistocene Transition. Despite the shift in the periodicity of the climate oscillations from 40,000 to 100,000 years, there was no shift in the underlying astronomical drivers—eccentricity and obliquity cycles were entirely the same before and after that transition.

Lest you wrongly conclude that this means that Milankovitch's theory of the ice ages is simply wrong, it's worth keeping in mind that Newton's theory of gravity—which provides a remarkably accurate description of everything from falling apples to the planetary orbits—is technically "wrong" in the same sense. It doesn't apply at very small scales (where we need quantum mechanics) or to massive objects like black holes (where we need general relativity). Science builds on itself, with more tentative, limited theories giving way to more robust, general theories. The Milankovitch theory, similarly, would require modifications to explain the MPT and the dominance of the 100,000-year sawtooth cycles.

Some, such as Brown University climatologist John Imbrie, recognized that the response to astronomical drivers was not a linear one. The relationship is more complicated, mediated through nonlinear amplifying mechanisms involving the behavior of ice sheets and the ocean carbon cycle.[21]

Others, however, boldly argued for abandoning the Milankovitch theory altogether. Physicist and astronomer Richard Muller is an iconoclast by any measure. He played the role of climate contrarian for a number of years, publicly chastising mainstream climate scientists for drawing a link between climate change and fossil fuel burning, only to claim in 2012, with much fanfare, to have proven the linkage himself. In 2016, Muller cofounded with his daughter Elizabeth a private for-profit company called Deep Isolation that promises to safely dispose of nuclear waste in deep holes in the ground. Its webpage features an image of Muller and his coworkers doing yoga in the desert.[22]

Back in the 1980s when I was doing my degree in physics at UC Berkeley, Muller was a faculty member in that department. I didn't take any of his classes, but I was aware of his work. He was a protégé of Luis Alvarez, the Nobel Prize–winning physicist who discovered the K-Pg asteroid impact that killed the dinosaurs. Perhaps inspired by the work of his adviser, Muller had searched for extraterrestrial explanations for other key events in Earth's history.

In 1984, the year I arrived at UC Berkeley, Muller had just proposed the "Death Star" hypothesis to explain a purported twenty-seven-million-year cyclicity in geological mass extinction events. The hypothesis was that a companion star to our Sun, a red dwarf star one and a half light years from Earth called Nemesis, traveled through the Oort cloud every twenty-seven million years, deflecting extra comets toward the inner solar system, where they might strike Earth and cause extinctions, analogous to the K-Pg impact event. The hypothesis has not held up. It is unclear that the periodicity actually exists in the mass extinction record in the first place, and there is no evidence of the hypothesized companion star.[23]

In the mid-1990s, as I was finishing up my Ph.D. at Yale University, Muller turned his gaze upon a different cycle—the 100,000-year ice age cycle. It came to be known as the Muller and MacDonald hypothesis (Muller's collaborator was Gordon MacDonald, a prominent geophysicist long known for his skepticism about plate tectonics). The hypothesis was that the 100,000-year periodicity had nothing to do at all with eccentricity. The shared timescale was just a coincidence. Instead, they argued that the ice age cycle was driven by a 100,000-year cyclicity in the inclination of the plane that defines Earth's orbit relative to the Sun. They hypothesized that the plane would, at some point during the course of this cycle, tilt into an interstellar dust cloud that would block some of the Sun's light from reaching Earth, cooling the planet slightly. If this sounds similar to Muller's earlier "Death Star" hypothesis, it's because it's similar to Muller's earlier "Death Star" hypothesis. And yes, this hypothesis would meet a similar fate.[24]

The arguments were unconvincing, and not backed up by rigorous and objective time series analysis methods. Extraordinary claims require extraordinary evidence. And Muller's arguments, in my view, didn't come close to meeting that burden. But don't take my word for it.

In 1996, prominent Columbia University geochemist and paleoclimatologist Wallace Broecker—himself somewhat of an iconoclast (he did, after all, title a book *The Glacial World According to Wally*)—entered the fray. Broecker largely adhered to the Milankovitch theory, but relished robust scientific debates. He gave Muller a hearing, holding a workshop at the Lamont-Doherty Earth Observatory, a scenic Columbia University–run research campus perched atop the cliffs of the Hudson River Palisades. Broecker invited the leading experts in the field to participate in a vigorous debate. It did not go well for Muller. Richard Kerr of *Science* magazine quoted leading experts like Dick Peltier, who said, "his arguments would just not sell," and time series analysis guru David Thomson of Bell Labs, who was "never impressed with that whole argument." Maureen Raymo cut to the quick, stating "orbital inclination has nothing to do with it." Broecker, the organizer of the workshop, was brutal in his assessment: "Even though you can't pin down exactly why Earth's climate responds to Milankovitch [orbital] cycles, at least there is some physical connection, whereas Rich Muller has none."[25]

My Ph.D. adviser Barry Saltzman was one of the participants in the workshop. Saltzman was a quiet, mild-mannered, self-effacing scientist—in many respects the polar opposite of Broecker. But his contributions to atmospheric science were substantial. It was Saltzman's research that led to the discovery, by MIT atmospheric scientist Ed Lorenz, of the so-called butterfly effect—that is, the nonlinear dynamics behind the weather.[26]

Saltzman's interest in nonlinear dynamical systems eventually led him to the problem of the Pleistocene glacial oscillations. Collaborating with other researchers including his former graduate student and post-doc Kirk Maasch (now a faculty member at the University of Maine) and former research associate Mikhail Verbitsky, Saltzman published a series of articles in the late 1980s and early 1990s demonstrating that the 100,000-year oscillation could be modeled as a free oscillation resulting from nonlinear mechanisms like ice bedrock depression and rebound, iceberg calving, ocean circulation, and carbon cycle dynamics, all paced by the Earth-orbital forcing. He showed that the system can exhibit a transition very much like the MPT from a predominantly 40,000-year cycle to an approximate 100,000-year cycle due to the long-term tectonically induced decline in CO_2 concentrations and the altered dynamics of the global climate system once a very large North American ice sheet emerges.[27]

Saltzman encountered resistance to his research findings from other prominent players, including Broecker. After returning from the 1996 workshop, he reported back to us with a combination of amusement and bemusement that Broecker, after having criticized his work for years, appeared to endorse and even take credit for it at the end of the workshop: "As I have believed all along, it appears that the best explanation for the [100,000-year] oscillation is that it is an internal oscillation paced by eccentricity."

And this remains our understanding today. Recent work by Matteo Willeit, Andrey Ganopolski, and collaborators at the Potsdam Institute for Climate Impact Research closely reproduces the climate record of the last three million years, including the MPT and the onset of sawtooth-shaped, 100,000-year cycles around a million years ago. They employed a climate model that is efficient enough that it can be run for millions of years, but comprehensive enough that it includes vital ice sheet and carbon cycle components. In a landmark 2019 study, the researchers simulated the way climate, ice sheets, and CO_2 levels from the late Pliocene evolved together through the Pleistocene, demonstrating that gradual lowering of atmospheric CO_2 and inclusion of what is known as regolith removal are essential to reproduce the observed climate history. Regolith is the layer of soil, dust, and loose rock that sits atop bedrock. It is easier for ice sheets to slide over this unconsolidated matter than the bedrock itself. With each successive glaciation, more and more regolith has been eroded away, increasingly exposing the bedrock. So, each successive glaciation encounters more friction with Earth's surface. With glaciers

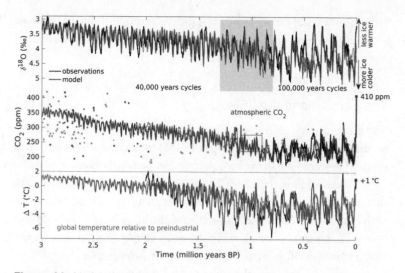

Figure 16. Model-simulated and observed variations in oxygen isotopes, CO_2, and global temperatures from late Pliocene through the present. The Mid-Pleistocene Transition is highlighted in gray.

stuck in place and unable to slide, it's much easier to build up larger and larger ice sheets.[28]

The researchers showed that the long-term, tectonically driven decrease in CO_2 leads to the initiation of Northern Hemisphere glaciation in the late Pliocene and an increase in the amplitude of glacial/interglacial cycles through the Pleistocene. The combined effect of the CO_2 decrease and gradual removal of regolith led to the transition from a smooth 40,000-year cycle and small ice sheets to a sawtooth-shaped, 100,000-year cycle and large ice sheets around the time of the actual MPT. In separate work, the group demonstrated that the sawtooth-shaped 100,000-year cycles result from "strongly nonlinear" responses to orbital forcing, synchronized with eccentricity. The "Glacial World According to *Barry*," it seems, was actually about right.[29]

We can now confidently say that the same models that reproduce the observed human-caused warming of the past century and project significant warming and dire consequences in the absence of dramatic reductions in carbon emissions can indeed reproduce one of the most striking phenomena in the paleoclimate record—the Pleistocene ice ages, including the once enigmatic MPT.

The Eemian and the LGM

The most recent ice age cycle began with an interglacial period known as the Eemian (around 130,000–115,000 years ago), followed by a gradual descent into the subsequent glacial period, culminating about 21,000 years ago in the Last Glacial Maximum (LGM). As we know, this was an especially important timeframe from the standpoint of human civilization. It was the time period during which *Homo sapiens* learned to collect and cook shellfish, make fishing tools, use language, and become, well, human. Anthropologists believe that the huge alteration in climate over this timeframe—a consequence, as we now know, of increasingly lower CO_2 and increasingly larger ice sheets— placed selective pressures on the very sort of intelligence that generally now characterizes our species. But this alteration was also significant from a climatic point of view. There are several important lessons we can take away from both the Last Interglacial (the Eemian), an interval which may have experienced even greater warming than today, and the LGM, when an ice sheet covered what is now New York City.

Let's start with the Eemian. Of all the past interglacial periods, it is the most relevant for it is the most recent and the best documented, and it affords us an opportunity to examine a period when CO_2 levels were similar to the preindustrial era (about 280 ppm), yet global temperatures were comparable to today. The geological evidence also indicates that global sea level was twenty to thirty feet higher than today, which would have required substantial melt contributions from both Antarctica (about ten to twenty-three feet) and Greenland (five to seven feet).[30]

Fossil evidence indicates that tropical fauna such as hippos and straight-tusked elephants were found in Europe as far north as the British Isles. Brown bears made it all the way up to the Alaskan Arctic, interbreeding with polar bears. Forests extended considerably farther north in North America and Eurasia, with trees found as far north as southern Baffin Island in the Canadian Arctic. Regions of Alaska that are tundra today were boreal forest. Then comes the question: How was such apparent warmth possible at CO_2 levels no higher than modern preindustrial levels?[31]

There was a confluence of natural factors that appear to have been at play. As with most interglacial periods, the Eemian coincided with a time of high orbital eccentricity, but it also coincided with high tilt of Earth's axis and precession that led to Earth's closest position relative to the sun coinciding with boreal summer. This combination of factors favored especially warm Arctic summers in particular, but also warm Antarctic summers, conditions that would have meant substantial melting of both the Greenland and West Antarctic Ice Sheets. Though average annual global temperatures may have been similar to today, summers at high northern latitudes may have been 3.5–7°F warmer.[32]

So, there's a bit of good news here. The substantial ice melt and sea level rise during the Eemian was a consequence of the unique Earth orbital configuration that favored unusual high-latitude summer warming. Similar levels of warming today can likely be avoided given meaningful efforts to reduce carbon emissions. A bit more good news: Both Arctic sea ice and polar bears appear to have survived the high-latitude warming of the Eemian, suggesting that they can survive human-caused climate change, if we can prevent substantial additional warming of the planet. There's no sign that any runaway Arctic permafrost "methane bomb" occurred, despite Arctic summers that were warmer than today.[33]

As Earth orbital factors began instead to favor decreased seasonality, cooler summers, and larger ice sheets, the climate slowly descended over the next hundred thousand years into the depths of the LGM. It was the largest swing yet in the 100,000-year sawtooth oscillations, resulting in the most extensive glaciation of all, with the most widespread ice extent occurring around 18,000–22,000 years ago. Levels of CO_2 dropped all the way down to about 180 ppm. How cold did it get? The most precise estimate to date is that the LGM was roughly 11°F cooler than the preindustrial era and roughly 13°F cooler than today. Cooling was considerably greater (about 36°F) in glaciated, high-latitude land regions than in tropical ocean regions (about 4°F).[34]

The LGM was a very different-looking world. The northern half of North America and Europe were covered by an ice sheet as much as two miles thick. The ice dome pushed the jet stream southward, and mid-latitude regions that are arid or semi-arid today were home then to huge inland lakes. The Utah salt flats are the remaining bed of what was 20,000-square-mile Lake Bonneville. I've viewed its ancient shoreline, visible above the valley floor from Salt Lake City. Because so much water was tied up in the ice sheets, global sea level was more than 400 feet lower than today.

Magnificent woolly mammoths and mastodons, giant ground sloths, and saber-toothed cats roamed North America and Europe. If you want a sense of what that world might have looked like, you needn't rely upon your imagination. The La Brea Tar Pits Museum in Los Angeles has re-created it for us. Specimens of each of these now-extinct megafauna at some point during the last ice age became trapped in the tar pits—sticky ponds of asphalt. That preserved them for their subsequent excavation and display. In fall 2017, I participated in a climate change forum at the museum, which is located at

Figure 17. Re-creation of a Pleistocene scene at the Lake Pit at La Brea Tar Pits.

the center of those ancient pools of asphalt, the viscous, evaporated remains of crude oil that seeped to the surface from deep below. It was hard not to see an ironic connection between the venue and the climate theme of the event. Crude oil from beneath Earth's surface threatens us today because we're ensnared by it politically rather than physically.[35]

The hundreds of feet of sea level lowering during the LGM, as you may recall, is what created the land bridge that allowed for the migration of *Homo sapiens* from Asia into North America. There remains an active debate over precisely when this happened. Was it the height of the LGM, 21,000 years ago? Or was it several thousand years later? Why did those iconic aforementioned megafauna, which had somehow survived all of the previous glacial/interglacial cycles (as well as the tar pits!), go extinct around this time? Was it hunting "overkill" by the newly arrived predators—us? Or was it the challenges of an unusually dramatic climate swing? That question remains hotly contested, as well.[36]

The LGM, in any event, presents a seemingly very attractive target for climate models. It provides, first of all, a big signal—a cooling of 11°F relative to the modern preindustrial level. The cooling is fairly precisely known, thanks to the relatively widespread proxy data available that far back. The climate remained that cold for several thousand years, meaning that it was in a quasi-equilibrium state. We also have reasonably reliable estimates of the climate drivers, including CO_2 concentrations from ice cores, and we have geological evidence that establishes the distribution of ice sheets. The marked contrast between unusually cold polar regions and the warm tropics during the LGM led to a strengthened jet stream, with lots of reflective, wind-blown dust in the atmosphere. We have good records of that, too, from ice cores. So, you might think, it's a cinch to back out a reliable estimate of climate sensitivity from the LGM or the Pleistocene glacial cycles more generally. But, in what is becoming a theme of this book, it's not that simple.

There is, first of all, a chicken-and-egg problem to deal with. It is perhaps best illustrated by a dust-up over the 2006 film *An Inconvenient Truth* featuring former U.S. vice president Al Gore. Second only to the Leonardo DiCaprio vehicle *Before The Flood* as the most viewed climate-themed documentary of all time, it has been credited for raising international public awareness of the climate crisis. It was also, predictably, the target of climate change deniers seeking to discredit the film, its message, and its messenger.[37]

Among the criticisms was that of an animation shown in the film comparing the Vostok records of CO_2 and temperature over the last 650,000 years, which, as we've already seen, vary in lockstep. Gore in the film characterizes that relationship as "an exact fit." We litigated this and other criticisms of the film on the climate scientist–run group blog *RealClimate* that I cofounded with Gavin Schmidt back in 2005. Our finding was that "Gore stated that the greenhouse gas levels and temperature changes over ice age signals had a complex relationship but that they 'fit'" and that "both of these statements are true." Gore's terse explanation leaves out some of that complexity.[38]

As we noted in the post, a full understanding of the CO_2 changes requires consideration of the additional so-called coke bottle effect: when changing Earth orbital factors favor warming over the southern ocean, the warm ocean loses CO_2 to the atmosphere in the same way that a warm coke bottle opened on a summer day quickly loses its CO_2 (eventually going flat!). The higher CO_2 then leads to further warming through the greenhouse effect—it's yet another positive feedback mechanism that helps produce the rapid warming and deglaciation observed as we transition from glacial to interglacial. The modeling work done by the Potsdam group discussed earlier demonstrates the importance of these mechanisms in reproducing the details of the observed relationship between the CO_2 and temperature curves. But the crux of Gore's point—that the observed, long-term relationship between CO_2 and temperature in Antarctica supports our understanding of the warming impact of increased CO_2 concentrations—is correct.

We already know that we can't estimate climate sensitivity using the Vostok temperature curve, as it shows twice as much warming as sediment-derived estimates of global temperature due to polar amplification of warming. But what the above complication means is that we cannot deduce the climate sensitivity by simply comparing the global CO_2 and temperature curves anyway. On this timescale, it's not a simple case of cause and effect. The level of CO_2 is not simply a lever arm controlling the climate system. Instead, it's one of the interacting variables, participating as both a *response to* and *cause of* warming and cooling.

This issue lies at the center of current disputes over what the Pleistocene ice ages tell us about the sensitivity of Earth's climate. A 2016 *Nature* study used precisely such a comparison of global temperature with CO_2 changes over the Pleistocene to infer a whopping 16°F value of ESS, arguing for an eventual warming of 3–7°C even if CO_2 were held at current levels. The study generated breathless media accounts, with *Pacific Standard* reporting that "doubling greenhouse gas emissions would warm the planet by about nine degrees [Celsius, 16°F]" and *Cosmos Magazine* warning us of "a grim global temperature rise of . . . as much as 7°C [13°F], if greenhouse gas emissions aren't reduced."[39]

The problem is that this just isn't correct. Twelve leading paleoclimatologists, including several whose work we've encountered in this chapter, published a response in *Nature* pointing out that one can't estimate climate sensitivity through a simple comparison of the two curves. They showed that the method fails when applied to a test case where the true ESS is known exactly beforehand. They reiterated that "there is no reason to alter the most recent assessment of the present-day committed warming." According to the latest IPCC report, that's about 3°F in a scenario of rapid decarbonization.[40]

A comprehensive recent assessment of the LGM attempted to match a climate model driven by state-of-the-art estimates of drivers, or forcings, including CO_2 and dust to the observations available back through the LGM to estimate climate sensitivity. The albedo from ice sheets was treated as a radiative driver rather than a slow response of the climate system, allowing them to diagnose a value of ECS, coming up with the range 4.3–8.1°F (central estimate of 6.1°F), well within the consensus range and contradicting the notion that the Pleistocene ice ages point toward an unusually high value of climate sensitivity. Some more good news. Not only can we explain the Pleistocene ice ages now, but we can show that they are consistent with other lines of evidence that suggest mid-range values of climate sensitivity. In other words, climate change is a threat, but it's still a manageable threat.[41]

Lessons Learned

Some of the most important lessons we can draw from the late Cenozoic icehouse regard the potential threat posed today by ice sheet collapse and sea level rise. We know that the Greenland Ice Sheet didn't come into being until the end of the Pliocene epoch or the beginning of the Pleistocene epoch, a little less than three million years ago. Carbon dioxide levels were likely lower than today. However, because of the phenomenon of hysteresis, the CO_2 level for losing the ice sheet is likely higher than the level for forming it, and at least slightly higher than the current level. That's little cause for comfort, however. Given all of the uncertainties, we could be very close to the warming threshold for losing the GIS, and we could also be close to the threshold for losing the marine portions of the WAIS and perhaps a bit of the EAIS as well. All told, that's around thirty-six feet of sea level rise we could commit to in the very near future in the absence of concerted climate action. Land that is home to a quarter of the current U.S. population and more than a billion people worldwide would be flooded.[42]

These numbers are sobering. However, they don't tell us how quickly this might happen. Centuries? Decades? The paleoclimate record, on its own, can't really give us those answers. Yet, the record doesn't exist in a vacuum. State-of-the-art ice sheet modeling and climate models provide us some important clues, and when combined with inferences from the paleoclimate record, as we will see shortly, they lead us to somewhat stark conclusions.

First, though, an important point about how science works: The vast majority of scientific studies—including most of those discussed in this book—advance our understanding incrementally. Very few lead to a paradigm shift—a fundamental alteration in how we think about a particular problem. There are exceptions, of course. These include the 1980 Alvarez and Alvarez article identifying the dinosaur-extinguishing Chicxulub impact event and the 1963 Lorenz article introducing the butterfly effect.

The 2016 *Nature* article by Rob DeConto and Dave Pollard, in my view, is one other example. It represented a significant enough development that Pollard, a Penn State colleague at the time, requested that the video stream for a seminar he was giving at our institute be cut before he described his latest findings, which were strictly under embargo by the journal.[43]

Prior to their study, the conventional wisdom—as reported, for example, by the IPCC in their fourth assessment report—was that a worst-case sea level rise scenario was roughly three feet by the end of the century. The DeConto and Pollard study single-handedly doubled that number to around six feet. That's enough sea level rise to displace more than 600 million people worldwide. As my friend Jeff Goodell, a *Rolling Stone* journalist covering the climate beat, puts it, "The difference between three feet and six feet is the difference between a manageable coastal evacuation and a decades-long refugee disaster. For many Pacific island nations, it is the difference between survival and extinction."[44]

Before we get into the DeConto and Pollard study, let's first revisit the so-called Marine Ice Sheet Instability or MISI (pronounced "micey") that we encountered earlier in this chapter, where retreating grounding lines, thinning ice shelves, and loss of buttressing can cause a runaway collapse of glaciers and eventually—possibly—an entire ice sheet.

We've known about MISI since the late 1970s. But it is no longer simply theory; it's reality. The tenuous marine glaciers of the WAIS hold enough ice to yield ten feet of sea level rise. Two in particular—the Thwaites and Pine Island Glaciers—are often called the "weak underbelly" of the WAIS, as their collapse would ensure the loss of a large chunk of the WAIS. Thwaites has been called the Doomsday Glacier because its collapse alone would produce more than two feet of sea level rise.[45]

A combination of modeling, satellite-based mapping, and gravitational measurements were used by NASA glaciologist Eric Rignot and colleagues to document grounding line retreat and ice mass

loss among these key glaciers, including an 8.7-mile retreat of the
Thwaites Glacier, over a twenty year timeframe. The authors omi-
nously concluded: "We find no major bed obstacle that would prevent
the glaciers from further retreat and draw down the entire basin." Ian
Joughin of the University of Washington employed a numerical model
of ice shelf dynamics to investigate the sensitivity of the Thwaites
Glacier to subsurface ocean melting. Joughin told Jeff Goodell that
"the process of marine ice-sheet destabilization is already underway
on Thwaites Glacier." He and colleagues have concluded that similar behavior may be underway for Pine Island Glacier. On the other
hand, they somewhat reassuringly predicted losses to be "moderate (less than a hundredth of an inch per year) over the 21st century" and concluded that irreversible collapse might not happen for
centuries.[46]

The novel factor that DeConto and Pollard introduced into the
calculations is what's instead known as the Marine Ice *Cliff* Instability
(MICI). It changes the picture substantially. The MICI (pronounced
"mickey") mechanism adds yet *another* amplifying feedback mechanism that has the potential to greatly accelerate ice sheet collapse. Because the submerged bedrock beneath the WAIS is depressed by the
massive load of ice above it, it slopes downward as you move away
from the coast toward the interior of the ice sheet. That means that
when an ice cliff face at the edge of a glacier collapses into the ocean,
it leaves an even taller one behind it. And when that ice cliff collapses,
an even taller one lies behind it. As the ice continues to erode, the ice
cliffs soon exceed the critical height of about 330 feet at which they
become unstable under their own weight and spontaneously collapse.
At that point, a runaway disintegration of the glacier can occur.

It is here that the late Cenozoic icehouse topic of this chapter
gets entwined in our story. DeConto and Pollard found that they
could not get their ice sheet model to reproduce the estimated ice
loss during the mid-Pliocene and Eemian unless they included the
MICI mechanism, the "structural collapse of marine-terminating ice
cliffs" as they put it in the article. Calibrating the model against Pliocene and Eemian sea level estimates, they then projected their model
forward from today under different future greenhouse gas emission
scenarios. A quantitative implementation, if you like, of the principle

that past is prologue. They found that Antarctica has the potential to contribute more than three feet of sea level rise by 2100 under business-as-usual carbon emissions. Adding that to the roughly three feet that had already been projected due to thermal expansion, melt of smaller glaciers around the world, and a bit of the Greenland Ice Sheet, yields about six feet of sea level rise by 2100. Under sustained carbon emissions beyond 2100, DeConto and Pollard found that surface melt due to atmospheric warming plays an increasing role, as does the partial collapse of the EAIS.[47]

This is all reason for concern, to be sure, and certainly the science here provides ample motivation for rapid societal decarbonization. But it does not spell doom. There is, in our public discourse over climate change these days, a hypersensitivity to seemingly every study that is published on ice sheet dynamics and sea level rise. The pattern is all too familiar to those of us in the climate communication space. The conclusions of the latest studies are hyped and climate doomists immediately point to them as purported confirmation of our unavoidable demise. There's plenty of blame to go around. Calling the Thwaites Glacier the Doomsday Glacier doesn't help, nor do breathless institutional press releases that overstate the significance of every *Science* or *Nature* article on the topic, insisting that their institution's latest research portends a revolutionary change in our thinking. Some studies are revolutionary. The Pollard and DeConto study is an example. But as mentioned earlier, very few studies are more than incremental in nature, and click-bait headlines often do an injustice to the more nuanced articles they purport to summarize.

As I was writing this chapter, a study had just been published in *Science* describing the detection of groundwater in sediments underlying the WAIS. An article on the CNN website warned: "Hidden deep below the ice sheet that covers Antarctica, scientists have discovered a massive amount of water." It quoted the lead author: "Antarctica contains 57 meters (187 feet) of sea level rise potential . . . Groundwater is currently a missing process in our models of ice flow." It's easy to see how a reader might conclude from this reporting that there is some huge missing physics in the models and we might be in store for much more ice melt and sea level rise than we thought.

As invariably happens in such instances, worried individuals reach out to me for a response. In this case, a Boulder-based climate solutions advocacy organization queried me via Twitter: "Does new mapping of a massive amount of groundwater below Antarctica's ice sheet have implications for climate change . . . ? Could a thinning ice sheet accelerate ice flows and sea level rise?" My go-to person on ice sheets is my former Penn State colleague Richard Alley, one of the leading experts in the field and a coauthor on several of the studies I've mentioned in this chapter. Alley, who is always extremely generous with his time, sent me a lengthy email describing why this was good science and an important contribution. But his bottom line? "I don't think [it is] very important for ice-sheet stability." Basically, ice sheet experts don't consider groundwater flow to be much of a factor in the behavior of large ice sheets. That's what I reported back to my interlocutor.[48]

The current scientific consensus is that we will continue to see melt from the Greenland and Antarctic ice sheets sustained at current rates for the remainder of this century, even in a scenario of substantial mitigation where we hold warming below 4°F. As one recent review article in the leading journal *Nature Climate Change* concludes, nonlinearities cannot be ruled out, so "large uncertainties in future projections still remain." That assessment found that if warming is maintained even at levels of 3–4°F for centuries, then, as alluded to before, we could cross key tipping points associated with the runaway collapse of the WAIS, driven by the Marine Ice Sheet and Ice Cliff Instabilities, and of the GIS, driven by the ice elevation feedback discussed earlier.[49]

Nonetheless, it could take many centuries for these processes to play out. A simultaneous pair of studies published in *Nature* in May 2021 making very different assumptions agree on this bottom-line conclusion. The more conservative of the two studies, led by Tamsin Edwards of King's College, London, skeptical of MICI, did not include it as a mechanism. The other, by Robert DeConto, David Pollard, and collaborators, did.[50]

It is important to explore the sensitivity to such assumptions. As my colleague Richard Alley told me, "some members of the community think that the chance of rapid collapse remains small . . . whereas other[s] . . . think that the chance becomes large . . . with larger and faster warming . . . There are good physical reasons why this uncertainty exists. Some of the uncertainty could be reduced . . . but some . . . may prove to be very difficult to reduce." The fact that both of these studies indicate that sea level rise can likely be kept to a manageable approximate three feet by the end of the century if carbon emissions are reduced rapidly and surface warming is held below 3°F is therefore significant—it suggests that even given the uncertainties, there's a good chance that we can still avoid catastrophic coastal inundation this century if we take concerted action.[51]

And there's a little bit of further good news here. DeConto and Pollard have recently refined their model, to account for a newly identified stabilizing feedback. When they include the effect of glacial runoff of meltwater into the ocean, they find that this leads to an expansion of sea ice along the periphery of Antarctica. That, in turn, leads to a cooling of the air and a decrease in surface melt, reducing the rate of ice sheet retreat through the end of this century.[52]

As a result of this refinement, DeConto and colleagues predict a minimal sea level contribution from Antarctic melt this century. The impact of the negative meltwater-runoff feedback does subside thereafter, however, and the Antarctic contribution from sea level rise in the subsequent century depends markedly on the carbon emissions scenario. If warming is kept below 4°F, the Antarctic sea level contribution remains below three feet by 2300. That's still likely around six feet total sea level rise, but with nearly two centuries intervening between now and then, it provides far more time to retreat from the encroaching seas.

Alternatively, under business-as-usual emissions with warming of 7–9°F this century, the contribution rises to a massive thirty-three feet. The authors warn that these findings "demonstrate the possibility that rapid and unstoppable sea-level rise from Antarctica will be triggered if Paris Agreement targets are exceeded." It would mean, among other things, the displacement of nearly a billion people.[53]

Buried in the ice is a message of both *urgency* and *agency*, and a little bit of reassurance regarding our fragile moment. We've seen that rapid reductions in carbon emissions can likely forestall collapsing ice sheets and massive sea level rise. We have also seen that we can avoid the triggering of hothouse feedbacks and regimes of enhanced climate sensitivity if we pursue meaningful climate action. However, failure to do so could lock in devastating, perhaps even civilization-ending, long-term impacts. It truly does appear to be up to us.

With that, we leave the Pleistocene epoch. But it is not the end of our climate journey. Quite a bit has happened, climatically speaking, since the last glacial cycle came to an end 12,000 years ago. Among other things, our own moment arose. Our final foray into the paleoclimate record will delve into that moment, zooming in on the last 2000 years, the Common Era, which includes both the baseline climate that existed prior to the industrial revolution and the last two centuries during which human beings emerged as the dominant driver of climate change. What lessons can we learn from this transitional period?

7

Beyond the Hockey Stick

> The hockey stick is now just one piece of a much
> broader picture of climate change . . . From the rapid
> melting of polar ice caps to the slowing of global ocean
> circulation, it's pretty clear that we're entering a new
> environmental era.
>
> —OLIVE HEFFERNAN, *The New Scientist*

More than two decades ago, when I was a post-doctoral re-searcher at the University of Massachusetts Amherst, I published the now famous hockey stick curve, along with my coauthors Raymond Bradley of UMass and Malcolm Hughes of the University of Arizona. It was a simple graph, derived from a global network of diverse climate proxy data such as tree rings, ice cores, corals, and lake sediments, depicting the average temperature of the Northern Hemisphere over the past millennium. The graph laid bare the unprecedented nature of the warming taking place today, and as a result it became a focal point in the debate over human-caused climate change and what to do about it.[1]

Yet, the apparent simplicity of the hockey stick curve betrays the dynamicism and complexity of the climate in past centuries and how it can inform our understanding of human-caused climate change and its impacts. In this chapter, we will see what other lessons we can learn from studying paleoclimate records and climate model simulations of the Common Era, the period of the past two millennia during which the signal of human-caused warming has risen dramatically above the noise of natural climate variability.

Clearly, there is a cautionary tale told by the hockey stick curve in the unprecedented warming that we are causing, but the lessons from the paleoclimate record of the Common Era go far beyond that. What might we learn, for example, from how key components of the climate system like the El Niño phenomenon and the Asian summer monsoon responded to natural drivers such as solar variations and volcanic eruptions? Are we nearing a potential tipping point for the North Atlantic Ocean conveyor belt circulation? How has sea level and tropical cyclone behavior varied over the past two millennia, and what might that tell us about future coastal risk? Are there natural, long-term oscillations, evident in the paleoclimate record, that might offer an alternative explanation of some recent trends? Can we get a better handle on climate sensitivity from examining how climate has responded to natural factors in the past? And, can better estimates of past climate trends inform assessments of how close we are to critical "dangerous" warming thresholds? In this chapter, we will seek answers to these questions through our examination of the Common Era.[2]

Enter the Holocene

Before we get to a discussion of the Common Era, a recap is in order. At the end of the previous chapter, we had come to the close of the Pleistocene. The ice melt began about 17,000 years ago, thanks to a favorable alignment of Earth orbital factors that led to warmer high-latitude summers. But we learned in Chapter 1 that there was a false start or two along the way to the Holocene, thanks to a flickering conveyor belt. And there were some other interesting climate responses as the glacial period took its last gasps.

When I was growing up in Amherst, Massachusetts, I would occasionally canoe down the Connecticut River in the summer, absorbing the bucolic scenery as the river winds its way through the fertile floodplain of the Connecticut River Valley. Cows would graze in the lush grass fields of quintessential New England farms, set against a background of the impossibly green mountain range known as the Seven

Sisters (which I frequently hiked). The floodplain was once filled by an ancient lake known as Glacial Lake Hitchcock, named after early American geologist (and third president of Amherst College) Edward Hitchcock, who first reported evidence of ancient lake deposits in the region back in 1818. The lake, at its maximum, extended from northern Vermont and New Hampshire all the way down to southern Connecticut.

Lake Hitchcock was actually a series of interconnected so-called proglacial lakes (lakes formed by the damming action of a terminal moraine or residual glacier) that were formed between 18,000 and 13,000 years ago, as the Laurentide Ice Sheet slowly retreated northward. The migrating terminus of the ice sheet produced torrents of meltwater, awash with glacial sediment that was flushed into the ever-northward-expanding lake. Just how much sediment piled up in any given year was determined by how warm the summer was and how much meltwater was generated. The seasonal alternation between summer melt and winter ice cover is visible as banding in the vertical layers of sediment left behind, known as varves. They can be counted back in time like the rings of a tree as one drills down into the sediment.

In the early 1920s, the Swedish-American geologist Ernst Antevs obtained sediment cores from various ancient Lake Hitchcock sites in Vermont and Connecticut, joining them together in an effort to create an approximate 4000-year series spanning the interval from 17,500 to 13,500 years before the present. But his original varve chronology had a couple gaps that remained unfilled. They were eventually filled in the late 1990s, when I was a post-doc in the Department of Geosciences at UMass, doing the research that led to the hockey stick curve. At that time, a Ph.D. student in the department named Tammy Rittenour was working with Julie Brigham-Grette, a leading glacial geologist and member of the faculty, to fill the final remaining gap. They had determined that the missing segment corresponded to the time period when the terminus of the ice sheet was passing through western Massachusetts. In fact, they pinpointed the optimal floodplain location to drill a core . . . the UMass soccer field! I like to imagine there was some amount of protest from the soccer coach, but they did get permission to drill the core. The result? The final piece of the puzzle and a continuous 4000-year record of summer temperatures in New England in the latter part of the last ice age.

I was known as a bit of a statistics guru in the department, and Rittenour and Brigham-Grette came to me for some assistance in analyzing this unique time series. Applying a statistical tool known as spectral analysis, we established that there were oscillations in the series with a periodicity of three to five years, consistent with the well-established timescale of the El Niño phenomenon. Originating in the tropical Pacific Ocean, El Niño today has a relatively weak downstream influence by the time you get all the way to New England. During the glacial period, however, when temperature contrasts between the equator and pole were considerably greater, the jet stream was much stronger, and El Niño's impact on distant regions was more extensive, likely reaching that far. We found that the amplitude of these oscillations decreased steadily over time, suggesting a weakening of El Niño as the climate transitioned from the Last Glacial Maximum to the end of the glacial period. The fact that climate change, whether natural or human-caused, can influence the internal dynamics of the climate system in this way is an important theme we have encountered before and will revisit later in this chapter.[3]

So, we finally exited the ice age around 12,000 years ago and entered the Holocene—the current interglacial period that serves as the baseline for how human-caused climate change is measured. The latest reconstructions of global temperature, as we saw in the introduction, show modest warming during the first few millennia as we continue to emerge from the last glacial period, and then an essentially flat global temperature trend once we reach the mid-Holocene, 6000 years ago. That doesn't mean that there weren't larger regional changes in climate. As we saw in Chapter 1, the roughly 26,000-year orbital precession cycle led to a decrease in seasonality and a weakening of monsoons from the early to mid-Holocene, leading to drier summers in subtropical regions and a resurgence of El Niño, which had gone dormant from the late glacial through the early Holocene. The paleoclimate evidence indicates gradually cooling northern summers. But averaged over the year and over the globe, temperature changes remained small over the past 6000 years, creating the remarkably stable global climate that led to our current habitable but fragile moment.[4]

As we have learned, some scientists, such as my former University of Virginia colleague William Ruddiman, have argued that the human impact on our climate might have begun well before the industrial revolution, more than 6000 years ago as human beings began to engage in agriculture and rice cultivation, deforestation, and other practices that generate greenhouse gases. This early greenhouse warming would have offset a slow, natural decline toward the next ice age.

Regardless of what one thinks of Ruddiman's early Anthropocene hypothesis (it does have its critics), it is clear we didn't *truly* take control of the climate lever arm until the dawn of the industrial revolution two centuries ago. It is then that we began to belch millions and then billions of tons of carbon pollution into the atmosphere through the burning of coal, oil, and fossil gas for energy and transportation, warming the planet and triggering a wide array of changes in our climate.[5]

Whether the Anthropocene began two centuries ago with the industrial revolution or five millennia ago with the development of widespread agriculture and land management, it is useful to zoom in on the Common Era, the period of the past two millennia during which modern human civilization developed. The preindustrial portion of this time period not only predates the impact of fossil fuel burning on our climate, but the basic "boundary conditions" of the climate—the geometry of Earth's orbit around the Sun, the extents of the ice sheets, and the distribution of vegetation—were largely unchanged from today. This time interval affords a natural control experiment—an opportunity to investigate what our climate *would* be like if not for the massive manner in which we are today altering our planetary environment.

The Hockey Stick

There is no better *measure* of the massive impact we are having on our climate today than the hockey stick curve. On the curve's twenty-year anniversary, April 22, 2018, *New Scientist* stated that "the hockey stick graph will always be climate science's icon." It has been featured in numerous books and in documentaries and films such as *An Inconvenient Truth*. It has been displayed on billboards and murals and even turned into music. It has also been the subject of attack by right-wing media outlets such as *Fox News* and *The Wall Street Journal*. It has generated congressional hearings and investigations in the United States and the United Kingdom by conservative politicians looking to discredit the graph. Why has so much attention been paid to a single graph published more than two decades ago?[6]

The hockey stick tells a simple story. You don't have to understand the complexities and intricacies of Earth's climate system to understand what this graph is telling us: that we are perturbing our planet's climate in a profound way. We know that there are times when the planet has been warmer. We only have to go back about 125,000 years to the penultimate interglacial—the Eemian—to find warmth that likely exceeded today's. But that warming happened over thousands of years, not a hundred years. Furthermore, there wasn't an entire civilization of nearly eight billion people whose existence was wholly dependent on climate stability back then. The warming spike that we are experiencing is unprecedented in the history of human civilization. The hockey stick communicates this simple truth in a stark and unforgiving manner.

Carl Sagan wrote eloquently about the "self-correcting machinery" of science—the peer review process, scientific assessments, and other checks and balances that keep science on the path toward a fuller and better understanding of the natural world. Findings that are wrong will be demonstrated so by other researchers. We saw a canonical example of that in Chapter 4 with our discussion of the pathological science of cold fusion. When findings are correct, on the other hand, they will be reproduced by other researchers. If validated over and over again, they become part of the accepted body of scientific understanding. Think plate tectonics, the laws of thermodynamics, or the theories of special and general relativity.[7]

The hockey stick was published as a sequence of two articles. The first, published in *Nature* in 1998, went back 600 years, and the second, published in *Geophysical Research Letters* a year later, extended

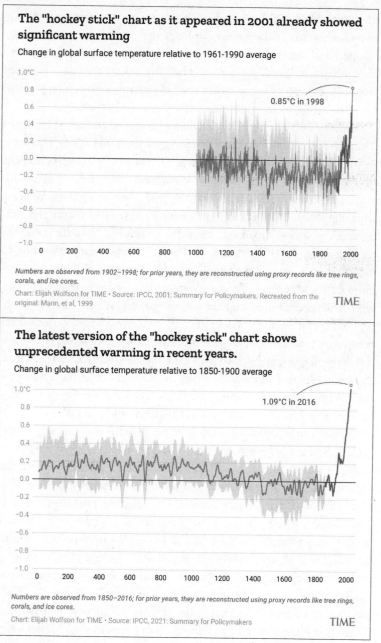

The "hockey stick" chart as it appeared in 2001 already showed significant warming

Change in global surface temperature relative to 1961-1990 average

0.85°C in 1998

Numbers are observed from 1902–1998; for prior years, they are reconstructed using proxy records like tree rings, corals, and ice cores.

Chart: Elijah Wolfson for TIME • Source: IPCC, 2001: Summary for Policymakers. Recreated from the original: Mann, et al, 1999

TIME

The latest version of the "hockey stick" chart shows unprecedented warming in recent years.

Change in global surface temperature relative to 1850-1900 average

1.09°C in 2016

Numbers are observed from 1850–2016; for prior years, they are reconstructed using proxy records like tree rings, corals, and ice cores.

Chart: Elijah Wolfson for TIME • Source: IPCC, 2021: Summary for Policymakers

TIME

Figure 18. Comparison of hockey stick curve reconstructions in the 2001 IPCC Third Assessment Report and 2021 IPCC Sixth Assessment Report.

the curve back 1000 years. Since then, far more paleoclimate data have become available, more-sophisticated methods have been developed and applied to these data, and longer reconstructions have been obtained. The net result is what I've referred to as a "hockey league"—dozens of independent studies that come to similar conclusions, resulting in an even longer, sturdier hockey stick.[8]

Back in 2001, the hockey stick was featured in the widely read "Summary for Policymakers" (or SPM) of the 2001 IPCC Third Assessment Report. Along with several other proxy reconstructions, it formed the basis of the tentative conclusion that the recent warming is unprecedented over at least the past 1000 years. Twenty years later, the 2021 IPCC Sixth Assessment Report SPM featured a longer hockey stick–shaped reconstruction, with the "handle" now extending back 2000 years and the "blade" rising even higher due to the continued warming of the past two decades (truth be told, it's more of a scythe than a hockey stick now—read into that what you will). The report found that the recent warming is very likely unprecedented for at least the past two millennia. And based on other recent, more tentative but longer reconstructions that extend back even further—all the way back to the LGM and beyond (such as the curve shown in Figure 1 in the introduction)—they concluded that the current warmth plausibly exceeds anything seen since the Eemian. That's more than a *hundred* millennia. Let that sink in.[9]

A critic might say this proves nothing. Perhaps the warming is natural, part of some long-term cycle? We'll get to the issue of long-term cycles in a bit, but there's a direct means of assessing what caused the anomalous warming. It involves what we call *detection and attribution* in the world of climate science. The detection part is easy in this case—we can clearly detect a sharp rise in temperature over the past century and a half relative to the preceding millennium. The attribution part is a bit more involved. It requires running climate models in two different scenarios: a counterfactual scenario where the industrial revolution didn't happen and no rise in carbon pollution took place, and a scenario reflecting the real-world history of fossil fuel burning. If a climate change–related event happens in the latter case, but not the former, we can *attribute* it to human activity.

Studies using climate models to simulate the two alternative scenarios demonstrate that natural factors, such as changes in the frequency and magnitude of volcanic eruptions or fluctuations in solar output, cannot reproduce the dramatic warming trend of the past century. In fact, the planet should have cooled slightly if only natural factors were at work. It is only when the human impacts—primarily the increase in CO_2 concentrations from fossil fuel burning, but secondarily the effect of atmospheric sulfur pollution—are included that the warming trend over the past century is reproduced. It's a simple fact: the planet is warming, and we are the cause.[10]

There is a cautionary tale told by the hockey stick curve when it comes to the unprecedented warming that we are causing through fossil fuel burning and other carbon-generating activities. But the lessons from the paleoclimate record of the Common Era go further. We can, for example, revisit the issue of assessing the sensitivity of the climate to ongoing human-caused increases in greenhouse gas concentrations. Examining how the climate responded to natural factors, such as volcanic eruptions and small but measurable changes in solar output in past centuries, informs our estimates of Equilibrium Climate Sensitivity. We can also gain insight into the behavior of various natural modes of climate variability and what they tell us about the dynamics of the climate system and the potential response of these modes to human-caused greenhouse forcing.

Thomas Jefferson, Climate Activist

Many of the key climate impacts involve so-called modes of variability of the climate system. The most familiar is the El Niño/Southern Oscillation (ENSO), which influences weather patterns around the world, impacting drought in the western United States, hurricane activity in the Atlantic and Pacific Basins, and flooding in East Africa and Australia, among many other impacts. You may recall the disastrous floods and mudslides in California during the winter of 2016. That was El Niño. Or the record-breaking thirty-one named storms of the 2020 Atlantic hurricane season? That was its flip side, La Niña.

Then there is the North Atlantic Oscillation (NAO), which we encountered back in Chapter 1, which impacts winter weather patterns in North America and Eurasia. And there's the Asian summer monsoon upon which more than a billion people in China, India, and other Asian countries depend today for their freshwater supply. We can better understand these key modes of variability of the climate system and their potential role in climate change by studying how they varied in the past.

There is no doubt, though, that ENSO is king among climate modes. It even competes with climate change itself for the title of the largest signal in the climate record. We already discussed the important role it plays with rainfall and drought in the western United States. However, it can also impact late spring and early summer rainfall and drought in the eastern United States, with dry conditions typically following a strong El Niño. In fact, a historically documented mega–El Niño during the early 1790s appears to have led to a massive drought in Virginia in spring 1792, described by James Madison in a letter to his friend Thomas Jefferson:

> I found this Country labouring under a most severe drought. There had been no rain whatever since the 18 or 20 of April. The flax and oats generally destroyed; The corn dying in the hills, no tobacco planted, and the wheat in weak land suffering.[11]

Back while I was a faculty member at the University of Virginia (UVA) in Charlottesville in the early 2000s, I participated in a tree-coring expedition with Columbia University and Lamont-Doherty Earth Observatory (LDEO) dendroclimatologist Edward Cook and his entire family. We set out to core old-growth trees from the grounds of Monticello, the onetime Charlottesville home of Thomas Jefferson, founder of UVA. We were well into the effort when the head groundskeeper approached rather hastily and aggressively. My imagination might be getting the better of me, but swear I remember him wielding an axe as he came toward us.

The right hand (the historical division that authorized our coring effort) apparently wasn't talking to the left hand (the groundskeeping division), so the head groundskeeper was unaware that this project had been sanctioned. He was, suffice it to say, rather upset. Ed explained why properly procured tree cores are not a threat to healthy trees, and that we were doing this with the blessing of the historical division. The purpose of the coring in this case was historical rather than paleoclimatic. Monticello historians were hoping to date some early nineteenth-century wooden historical structures they had excavated by matching the rings from the ancient wood against a chronology established from the tree cores. We escaped unharmed with our tree cores intact.

I would later participate in a similar tree-coring expedition whose impetus *was* paleoclimatic. It was a couple years later at Montpelier, James Madison's home about twenty miles northeast of Charlottesville. A UVA graduate student Daniel Druckenbrod (now chair of the Department of Geological, Environmental, and Marine Sciences at Rider University in New Jersey) had been coring trees as part of his Ph.D. project with my UVA colleague Hank Shugart, an expert on the modeling of forest ecosystems. They were interested in the age and size distribution of trees in this relatively undisturbed old-growth stand of trees to help with their modeling of forest dynamics. But Druckenbrod and I got to talking, and realized there was something else we might be able to do if we obtained some longer tree cores. They sometimes say that "one person's signal is another person's noise," and that was certainly the case here. The climate variability recorded by these trees partially masked the simple age-related growth trends they were trying to model. Those trends were their "signal," while the climate-induced variability about the trend was their "noise." Their noise was therefore *my* signal.

As it happens, I had been employing a couple UVA undergraduates to transcribe Jefferson's weather diaries from Monticello. He maintained meticulous records of temperature, rainfall, wind direction, and other such meteorological information, as well as a "garden book" in which he recorded the timing of phenological indicators—the arrival of bird species in the spring, the timing of flowering of plants, etc. Unfortunately, Jefferson frequently carried his meteorological instruments with him when traveling. As a result, his observations would shift from Monticello to other locations he visited such as Philadelphia, Annapolis, and Paris. However, his protégé Madison, at Jefferson's behest, assiduously maintained his own weather diaries at the fixed location of Montpelier, providing a consistent historical weather record from 1784 to 1802. Conveniently, that documentary record was colocated with the old-growth trees Druckenbrod had been coring.

We ended up collaborating with University of Arkansas tree ring expert David Stahle and colleagues to develop a dendroclimatic reconstruction from the tree ring data that we could compare with the historical records. Based on correlations with modern instrumental data, we found that the thickness of the rings in the latter half of the growth season closely tracked late spring/early summer rainfall in central Virginia. The tree ring–based rainfall reconstruction pointed to an unusually pronounced drought in 1792, following the massive 1791 El Niño event that is documented in historical reconstructions of past El Niño behavior. That's the very same drought that Madison mentioned in his letter to Jefferson. It was evident, too, in the very low May precipitation total for that year derived from Madison's weather diaries.[12]

A comparison of the eighteenth-century and early nineteenth-century Madison weather diary information with modern rain gauge data yielded another interesting finding: there has been a shift in the seasonal cycle of rainfall in the region, with the summer rainfall peak occurring more than a month later today than during Jefferson and Madison's time. Climate change reveals itself once again, likely a consequence of how it is nudging the timing of the seasons. Indeed, Jefferson himself, a brilliant amateur scientist of his time, was cognizant of the possibility that human activity could influence climate. He was unaware of the greenhouse effect (he died in 1826, only two years after it was first postulated by Joseph Fourier, a scientific development he understandably might not have stayed up on). Instead, he imagined it was deforestation and land clearing that would impact the climate:

> Years are requisite for this, steady attention to the thermometer, to
> the plants growing there, the times of their leafing and flowering, its
> animal inhabitants, beasts, birds, reptiles, and insects; its prevalent
> winds, quantities of rain and snow, temperature of fountains, and

other indexes of climate. We want this indeed for all the States, and
the work should be repeated once or twice in a century, to show the
effect of clearing and culture towards changes of climate.

You read that last line right: it's Thomas Jefferson writing about
"changes of climate" back in the early nineteenth century.[13]

But let us return to El Niño and climate change, and how the
paleoclimate record of the Common Era can inform our understanding of it. There is still a very open question, in fact, as to whether
human-caused greenhouse warming will lead to a more El Niño–like
world or the opposite, a La Niña–like world with relatively cool conditions in the eastern equatorial Pacific Ocean. The question is hardly
just academic. Drought in the desert Southwest, which impacts large
population centers in California, Nevada, and Arizona, is modulated
by ENSO. Though the influence varies from one event to the next, El
Niño years tend to be wetter than normal, and La Niña years tend
to be drier than normal. ENSO also impacts Atlantic hurricane activity, with El Niño years less active and La Niña years more active
than normal. Any changes in ENSO, including both its average state
and the magnitude of individual El Niño and La Niña events, could
have profound impacts around the world including North and South
America, Africa, Australia, and Indonesia.

Even state-of-the-art climate models, however, provide limited
guidance. They display a wide range in the response of the tropical
Pacific climate to human greenhouse gas heating. Moreover, the models, on the whole, are inconsistent with the observations. They exhibit
a trend toward an El Niño–like state, with a decreased contrast between the western tropical Pacific warm pool and the eastern tropical
Pacific "cold tongue"—a region of relatively cool water—whereas
the observations show a neutral or even opposite, La Niña–like trend
over the past half century. Leading LDEO oceanographer Richard
Seager and colleagues argue that the failure of many of the models to
reproduce the observed trend may arise from the fact that the upwelling of cold, deep water in the eastern equatorial Pacific is too strong
for the mechanisms governing ENSO to suppress.[14]

Seager's LDEO colleague Mark Cane—who created the first model
back in the late 1980s that could reproduce and predict El Niño

events—coauthored an article in 1997 arguing that the trade wind–induced upwelling of cold waters in the eastern equatorial Pacific opposes the warming influence of greenhouse gases, while the western equatorial Pacific remains free to warm. The so-called Bjerknes feedbacks govern the relationship between trade winds and the east-west variation in sea surface temperatures (SSTs), and are responsible for ENSO in the first place. Those feedbacks reinforce the initial response, creating even stronger trade winds and a stronger east-west difference in SSTs, with a warmer western equatorial Pacific and a colder eastern equatorial Pacific—an overall La Niña–like pattern![15]

Some recent work confirms this mechanism, but provides a more nuanced picture, indicating a transient tug-of-war between the direct warming impact of greenhouse gases (which favors an El Niño–like pattern) and the dynamical response of the Bjerknes feedbacks to that warming (which favors a La Niña–like pattern). Which of these competing effects wins out may depend on how long a timeframe we are looking at. The latter, dynamical mechanism appears to be more relevant on multidecadal to centennial timescales. And that's where paleoclimate evidence sheds considerable light.[16]

We can gain insight into the potential role of this mechanism in current-day greenhouse warming by looking at how ENSO responded to natural drivers in the past, such as changes in solar output and volcanic eruptions. Proxy data spanning the past millennium, such as tree rings, corals, lake sediments, and ice cores, appear consistent with a La Niña–like cooling in the eastern equatorial Pacific and a pattern in the United States of a dry desert Southwest and wet Pacific Northwest during the early part of the past millennium (1000 to 1400 CE) that is consistent with a La Niña–like state. Accordant with the mechanism of Mark Cane and colleagues, that state coincides with a period of external heating with high solar output and few explosive tropical volcanic eruptions. A number of studies find a tendency for an El Niño–like response to volcanic cooling events, once again consistent with Cane et al.'s mechanism. This topic continues to be debated in the literature.[17]

The medieval La Niña pattern is not in general reproduced in global coupled climate model simulations, which could be due to biases in the models, including the overly strong upwelling noted

earlier and possible limitations in the response of marine stratocumulus clouds to warming—a problem alluded to in our discussion
of past hothouse climates. The fact that the models aren't capturing
this apparent paleoclimatic response to natural drivers provides additional support for the hypothesis that they may not be capturing a La
Niña–like response to greenhouse warming. If so, they are *underestimating* key climate change impacts such as the heightened hurricane
activity in the tropical Atlantic Ocean and the growing aridification
of the western United States. The desert Southwest is currently experiencing a more-than-decade-long drought that is unprecedented
in at least 1200 years and perhaps longer, according to the tree ring
evidence analyzed by my former student LDEO scientist Benjamin
Cook and colleagues, underscoring the importance of getting these
things right if we are to use climate model projections to prepare for
the challenges to come.[18]

Another important dynamical response to heating involves the
North Atlantic Oscillation, or NAO (or the closely related Arctic
Oscillation, or AO), discussed back in Chapter 1. These modes of
variation describe changes in the winter storm track from year to
year that are especially prominent over the North Atlantic sector and
impact winter temperatures and precipitation over a large part of
North America and Eurasia by redirecting the paths of storms. The
bitter cold and snowy winter of 2009/2010 in Europe and eastern
North America was associated with an exceptionally strong negative
NAO pattern. The negative phase of the AO/NAO is associated with
a weaker jet stream that travels straight across the North Atlantic
Ocean rather than veering northward toward Europe as it often does.
Cold Arctic air gets locked up in Europe, eastern North America, and
parts of Asia.

My own analyses of climate model simulations and paleoclimate
data suggest that the especially cold winters in Europe in the mid-
seventeenth century were associated with a strong negative NAO pattern at that time. The relative warmth of those same regions during
the medieval era, we found, was tied to the reverse pattern—the positive phase of the AO/NAO. Because the Little Ice Age was a period
of relatively cool global temperatures and the medieval era a period
of relative global warmth, it might be tempting for us to generalize

this relationship to current human-caused warming. But we would be wrong.[19]

These past NAO shifts appear to have been caused by changes in solar output and ultraviolet radiation, which influence ozone levels and the absorption of solar heating by the stratosphere, altering the vertical contrasts in atmospheric temperature that drive the jet stream. The bottom line? First, the few climate model simulations that include ozone photochemistry (which is critical to this mechanism) reproduce the surface temperature patterns that we see in proxy reconstructions, and models that lack these processes do not. This underscores another potential limitation in current climate models—important to keep in mind because, as we've learned time and again, uncertainty is often not our friend when it comes to climate predictions. Second, the responses were rather specific to the factor that was at play—changing solar output. That's different from the factor (greenhouse gas increases) behind current warming. This brings up another important caveat regarding paleoclimate episodes: they're not always appropriate analogs for what is happening today. It's important to know when they are and when they aren't, something I've tried to drive home throughout this book.

Last but not least is the South Asian summer monsoon (SASM). Monsoonal rainfall provides freshwater for a large population of more than a billion in South Asia, making its potential future behavior under climate change a matter of special importance. We can learn something about this important component of the climate system, once again, from studying its response to past natural drivers.

The SASM is characterized by rising motion over the Indian subcontinent driven by the contrast in solar heating in the summer between the fast-warming land regions and the more thermally sluggish Indian Ocean. It's analogous in some ways to a sea breeze circulation—a phenomenon that is familiar to seaside beach vacationers everywhere: late afternoon comes around and those blue skies turn threatening, with towering dark clouds, thunderstorms, and torrential rains. It's such a regular pattern that you can set your daily routine by it. The daytime sun heats up the land much faster than the sluggish ocean, building up a contrast in heating that drives an onshore circulation, with moist air coming off the ocean, lifting over the heated land, rising, cooling, condensing into raindrops that fall, and then descending back over the ocean to complete the circulation. After the sun goes down and the heating contrast disappears, the circulation weakens. A summer monsoon is basically the same idea, but the winter/summer contrast in heating plays the role of the day/night cycle, and the circulation is continental in scale, so the atmospheric Coriolis effect becomes important. (Without getting into the gruesome atmospheric physics, the Coriolis effect, which is due to the rotation of the Earth, causes winds to spiral in as they rise, for example, over the heated Tibetan plateau and to spiral out as they descend, for example, over the Indian Ocean.)

Air is lifted up as it spirals inland up the slopes of the Himalayas. We get dramatic towering cumulonimbus clouds that are so energized by the heat released from the condensing rising moisture that they can overshoot the boundary—known as the tropopause—between the relatively unstable troposphere (the first six to nine miles of the atmosphere where weather takes place) and the very stable stratosphere (the region of the atmosphere, lying above the troposphere, where jets fly). The SASM characterizes the atmospheric circulation, but it is sometimes equated with the precipitation that results from it. That's wrong.

To understand why, let's consider what's known as the wind-precipitation paradox. The paradox has to do with how the increase in moisture in the atmosphere from greenhouse warming leads to increased rainfall without a strengthened monsoonal circulation. The heat released from the condensation of that extra moisture warms the mid-levels of the troposphere, creating a more stable atmosphere (warm air on top of cold air is a stable configuration). That increased stability inhibits the rising motion needed for a monsoonal circulation. It is thus possible to have strong monsoonal circulation with no rain or monsoon rain with weak circulation, something that is seen in both historical simulations and in future projections from climate models, where monsoonal rains increase with warming while the circulation itself weakens. That's good news from the standpoint of water resources but bad news from the standpoint of flooding.[20]

What insights, then, can we derive from the behavior of the SASM in past centuries? In an analysis of a climate model simulation of the past millennium led by my former graduate student Fangxing Fan, we found a general similarity between the variations in SASM circulation and SASM rainfall during the preindustrial period. However, we witnessed a marked decoupling of the two during the modern era, when the SASM circulation weakens but SASM precipitation does not. This finding underscores again a key point: Relationships that may have held in the past don't necessarily hold today and won't necessarily hold in the future. We have to understand the science well enough to know when they do and when they don't.[21]

Ocean Conveyor Collapse

We return once again to the ocean conveyor belt, known more technically as the Atlantic Meridional Overturning Circulation (AMOC). The ribbon-like current system is driven by the sinking of cold, salty water in Baffin Bay, nestled between Greenland and northern Canada, and the Labrador and Norwegian-Greenland Seas. It delivers warm subtropical waters to the high latitudes of the North Atlantic, thereby keeping the North Atlantic and neighboring regions in North America and Europe warmer than they would otherwise be. The AMOC is tied to convective overturning, wherein cold, nutrient- and oxygen-rich waters from below mix into the upper ocean where they are otherwise constantly depleted by marine biota.

We have already discussed examples of tipping points, particularly the collapse of the ice sheets. But the collapse of the ocean conveyor is also an example of a tipping point element in the climate system. Once it happens, there may be no bringing it back, at least on societal timescales. Tipping points force us to confront the tenuousness of climate stability, for we do not know precisely where they lie. They are like mines in a minefield, and once we trigger them, it is too late to undo the damage. Could the collapse of the ocean conveyor be among the tipping points that threaten our fragile moment?

Such a scenario, as we already know, was caricatured in the movie *The Day After Tomorrow*. Though the science depicted in the film— tornadoes destroying Los Angeles, and an ice sheet re-forming over

North America in a matter of days—is terribly far-fetched, there *would* be far-reaching consequences. Among them are diminished fish populations in the North Atlantic Ocean, the world's most productive natural fishery; fiercer winter storms in Europe; accelerated sea level rise along parts of the U.S. East Coast (a side effect of the physics that governs ocean current systems); and the potential for increased Atlantic hurricane activity as heat is bottled up in the tropical North Atlantic.[22]

We have already seen how the paleoclimate record can inform our understanding of prospects for AMOC collapse. Back in Chapter 1 we discussed the role of freshwater input and AMOC weakening during the Younger-Dryas event as the last glacial period came to a close and the analogous, albeit more muted, 8200 BP event during the early Holocene. Might the Common Era have something to say about AMOC collapse today?

Climate models predict that the AMOC will weaken later *this century* if we continue burning carbon and warming the planet, due primarily to the meltwater from the Greenland Ice Sheet as it begins to disintegrate. The paleo observations spanning the past two millennia beg to differ, however—they suggest that a dramatic slowdown has *already* begun.

Stefan Rahmstorf of the Potsdam Institute for Climate Impact Research and collaborators (including myself) estimated changes in the AMOC over the course of the Common Era in a 2015 article in *Nature Climate Change*. We made use of two complementary sources of proxy information dating back to 500 CE. The first of these was a proxy reconstruction of North Atlantic surface temperatures in the "cold blob" region just south of Greenland that would be most influenced by an AMOC slowdown. The second was a proxy record of nitrogen isotopes from deep-sea corals in the Northwest Atlantic Ocean off the coast of Nova Scotia, which record changes in North Atlantic slope waters. A more recent study used marine sediment silt data and planktic and benthic foram proxy data that extend back to 400 CE.[23]

All of these paleoclimate data point to the same unsettling conclusion: There has been a slowdown in the AMOC over the past century that is unprecedented in the Common Era. Though climate models do not predict a substantial weakening of the AMOC until later this century, that weakening appears to have occurred already. Once again, we see that the models, far from being "alarmist" as climate contrarians are wont to insist, are in some respects actually overly conservative, sometimes underpredicting key impacts. Inevitably, some of the more fundamental scientific uncertainties are unlikely to be definitively resolved on the timeframe we might like, namely the next few years during which we have to make critical decisions about climate policy. We must fall back, once again, on the evergreen principle that uncertainty is not our friend. It is a reason for more, not less, urgent action.

What might the models be missing or not getting right in this particular case? The leading candidate is Greenland, specifically the melt of the GIS and the freshwater input into the North Atlantic Ocean from it. Greenland melt appears to be exceeding past model predictions. During the month of July 2019 alone, nearly 200 billion tons of meltwater drained into the North Atlantic Ocean—enough to raise global sea levels by a visually measurable, if small, amount (half millimeter). All of that glacial meltwater is freshening the North Atlantic, and doing so ahead of schedule.[24]

Though direct observations of the AMOC show somewhat conflicting trends in recent decades, the most comprehensive recent study, which compares a variety of complementary metrics, argues that AMOC collapse is underway. One current limitation of the climate projections that form the primary basis of the IPCC assessments is that the climate models are not typically coupled with comprehensive ice sheet models, so the interactions between ice sheet disintegration and melt, freshwater runoff, and AMOC dynamics is not well represented in the modeling experiments. It's another example of how these climate projections are in some respects too conservative, failing to resolve the full range of dynamics and coupling that exist in the actual climate system.[25]

That's an important caveat because there is increasing evidence that many of these responses interact with each other, implying potential cascading effects. As we learned earlier, there is evidence that greenhouse warming could push us into a more La Niña–like climate state. We have also just seen that climate change may be causing a weakening of the AMOC. There is now evidence that these responses might mutually reinforce each other. One recent study finds that AMOC collapse, by bottling up heat in the tropical Atlantic Ocean, leads to warmer sea surface temperatures in the tropical South Atlantic. This drives a large-scale, sea breeze–style tropical atmospheric circulation that strengthens the trade winds and increases the upwelling of cold, deep waters, yielding a more La Niña–like state.[26]

Lest one conclude that this stuff is overly academic or theoretical, consider what we've witnessed over the past few years. The oceans have continued to set records for global warmth year after year. During this same time period, however, both the "cold blob" region in the subpolar North Atlantic and the eastern tropical Pacific have seen record cold. And while we're on the topic of coupled responses and tipping points, it is worth keeping in mind that the earlier-than-expected Greenland Ice Sheet melt, which has contributed to a slowing AMOC, is also contributing to sea level rise earlier than expected. This brings us to our next topic: sea level rise, intensified tropical cyclones, and increased coastal risk.[27]

Threatened Coastlines

Climate change poses twin coastal threats in the form of sea level rise and more intense tropical cyclones. Studies of the Common Era inform our understanding of both phenomena, by establishing a baseline that we can use to evaluate the rate and magnitude of current changes and by elucidating the climate factors that have governed past changes.

A decade and a half ago, projections of future sea level rise were hampered by the limitations in climate modeling. The future sea level rise projections presented by the IPCC in their 2007 Fourth Assessment Report, for example, presented a worst-case-scenario rise of only a foot and a half by the end of the century, albeit noting the (serious!) caveat that they had left out contributions from ice sheets because of a lack of confidence in the ability to model them. These decisions, in turn, led to media coverage that vastly understated the true likely prospects for sea level rise in the decades ahead.[28]

The fact that the likely most important contribution to future sea level rise was left out of the definitive international climate assessment proved troubling from both a scientific and public communication standpoint. My colleague Stefan Rahmstorf saw the need to fill the void by introducing an alternative to process-based modeling, a so-called semi-empirical approach to projecting future sea level rise. The approach is based on the principle that the processes underlying anomalous sea level rise—including thermal expansion of seawater, melting of glaciers, and ice sheet collapse—can be represented by the warming relative to the preindustrial baseline.

In a high-profile 2007 article in the journal *Science*, Rahmstorf demonstrated a very close historical relationship between that warming and the rate of change in sea level during the historical era of the past century and a half. Using that relationship to project forward, he predicted as much as five feet of sea level rise by the end of the century under a worst-case emissions/warming scenario. Some experts remained skeptical of this novel approach. Might paleoclimate once again come to the rescue?[29]

Sea level during the Common Era can be estimated from coastal deposits. Back in 2006, I was part of a project led by coastal sedimentologist Andrew Kemp, now at Tufts University, using sediments from salt marshes along the U.S. East Coast to reconstruct the history of sea level rise over the Common Era. The tidal inundation of salt marshes, recorded by the sediments, is tied to global sea level. Certain types of forams are known to live at different levels of the water column. The foram fossils found in the sediments can thus be used to estimate local sea level back in time. Accounting for the isostatic rebound effects mentioned in Chapter 6, it is possible to isolate the component of regional sea level change that is due to global sea level rise. It is found to be similar among numerous independent sites, suggesting a robust, common, global sea level rise signal.[30]

The resulting record indicates that the current rate of sea level rise is unprecedented over the past two millennia, a conclusion that has been reaffirmed by more recent work. This is yet another measure

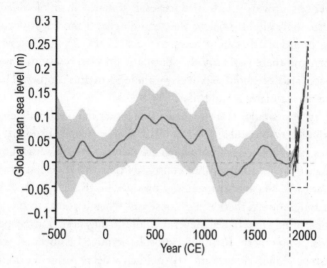

Figure 19. Global sea level rise over time.

of the profound impact of current human-caused planetary warming. The study went beyond that, however. It used a reconstruction of global temperatures spanning the past two millennia based on climate proxy data to drive Rahmstorf's semi-empirical model of global sea level rise. This alternative estimate reaffirmed the main features of the sediment-based sea level reconstruction, including the unprecedented nature of the recent increase, tying that increase to the unprecedented warming of the past century. It also provided an opportunity for further refinement of the statistical parameters of the semi-empirical model. Most importantly of all, it validated that model as a predictor of global sea level rise.[31]

Fast-forward six years, to the next IPCC (Fifth Assessment) report in 2013. Rahmstorf's semi-empirical model gets short shrift, reflecting ongoing skepticism about that approach on the part of the lead authors of that chapter. However, it would seem that the message of that work was heard. Thanks to the inclusion, for the first time, of the contribution from ice sheet dynamics (as represented instead by process modeling), the IPCC dramatically revised their sea level projections upward, with a worst-case scenario now of four feet, which is only slightly below the semi-empirical model projections. Fast-forward another eight years, to the 2021 IPCC Sixth Assessment Report, and there is yet another dramatic upward revision: a worst-case scenario of about 6.5 feet by the end of the century. That is where the estimates stand today.[32]

There is a similar story regarding the other major climate change contributor to coastal risk—tropical cyclones (tropical storms, hurricanes, and typhoons, all of which are members of the same family). Though there is an emerging consensus that the strongest tropical cyclones will become stronger and produce more damage from both wind and flooding, there is less consensus when it comes to the *number* of storms we will see. Several studies using nested atmospheric models (a fine-mesh, regional atmospheric model embedded within a coarser, global climate model) predict a decrease in the number of storms both globally and in the Atlantic Basin. But the higher-resolution model is often run at a resolution that is insufficient to resolve some important atmospheric processes. An alternative down-scaling approach used by Kerry Emanuel of MIT that avoids these problems (but does make some simplifying assumptions) comes to a very different conclusion, deducing from the most recent IPCC climate model simulations a substantial increase in both the intensity and number of storms in all basins.[33]

Paleoclimate observations and modeling of past tropical cyclone behavior can once again inform the discussion. The endeavor even has a name—paleotempestology—coined by Emanuel. Back in 2009, I collaborated with two paleotempestologists, Jonathan Woodruff of UMass Amherst and Jeff Donnelly of the Woods Hole Oceano-graphic Institution. They had spent years recovering valuable records of past hurricanes making landfall in the Caribbean and along the Gulf of Mexico and U.S. East Coast, derived once again from coastal sediments, albeit a very special type known as overwash deposits. These are ocean sediments that are found in coastal lagoons. They don't belong there. The only way they could have *gotten* there is by being transported by an extremely strong coastal storm (or possibly a tsunami, though these can often be distinguished based on other historical information). By drilling a sediment core down through the lagoon and using appropriate dating markers, one can recover a history of these storms.

We formed an Atlantic Basin–wide composite from the various overwash deposit records, yielding a history of major hurricanes that made landfall and impacted the Atlantic Basin going back 2000 years. We made separate use of a statistical model that predicts tropical cyclone activity based on climate variables such as tropical Atlantic sea temperatures and El Niño. Our statistical model has been used successfully to predict seasonal storm totals in advance of the season for a number of years—often outperforming other models. Driving the statistical model instead with reconstructions of those climate variables from proxy data, we were able to obtain an independent estimate of Atlantic tropical cyclone activity 1500 years back in time.[34]

We showed that both estimates—the sediment record of hurricanes making landfall and the statistical model driven by paleoclimate reconstructions—yield very similar histories. Both point to a period of high activity during the medieval era, which the semi-empirical model attributes to a combination of favorable factors—warm sea surface temperatures in the tropical Atlantic and La Niña conditions in the tropical Pacific. Tropical Atlantic warmth provides favorable thermodynamic conditions for tropical cyclone formation, while La Niña conditions are associated with diminished vertical wind shear, providing a more favorable atmospheric environment for tropical cyclones to form. Basin-wide Atlantic tropical cyclone activity over the past few decades rivals that of any comparable period during the entire record, attributed to this same combination of aggravating factors. If climate change indeed favors not only a warming tropical Atlantic but, as we've alluded to, a more La Niña–like climate, these findings portend a "perfect storm" of conditions favoring active Atlantic hurricane seasons.[35]

Meanwhile, rising sea level and more intense hurricanes pose an even greater coastal threat in combination. My former graduate student Andra Garner, now a faculty member at Rowan University in New Jersey, looked into the combined flood risk to New York City for her Ph.D. research. She used synthetic tropical cyclones based on Kerry Emanuel's downscaling approach applied to climate model simulations of the past millennium and the revised sea level estimates from the work of DeConto and Pollard discussed in the last chapter. Garner found that a seven-plus-foot flood, which would have been a 500-year flood (a flood that happens on average only once in 500 years) prior to human-caused global warming, has now become a roughly twenty-four-year flood, and will become a roughly five-year flood by mid-century under business-as-usual carbon emissions. Superstorm Sandy came with a seventy-billion-dollar price tag (at least eight billion of that attributable to climate change) when it struck New York and the Jersey coast in October 2012. Imagine New York City (and many other coastal cities) having to deal with something like this once every few years. That's where we're headed in the absence of concerted climate action.[36]

The Rise and Fall of the AMO

One argument contrarians often use to dispute the climate crisis is that many of the key trends are supposedly not tied to human-caused warming but are instead part of some low-frequency oscillation in the climate. If the rash of devastating Atlantic hurricanes that we've witnessed in recent years is just part of some long-term cycle, then why worry? We can just wait it out, right? It's a seemingly compelling claim, but it's not actually supported by the evidence. At all.

The analysis of paleoclimate data that led to the hockey stick reconstruction was in fact an outgrowth of my earlier Ph.D. work analyzing networks of proxy data to assess evidence of such natural long-term climate oscillations. In the early 1990s, statistical analyses of climate data in the North Atlantic Ocean and neighboring regions seemed to show evidence of a low-frequency oscillation, with warming through the mid-twentieth century, cooling through the 1970s, and warming thereafter. There was also some limited evidence emerging from the first long simulations of climate models that had dynamic rather than static oceans. Some of these coupled ocean-atmosphere model simulations produced long-term oscillations with a multidecadal fifty-to-seventy-year timescale. Like the interannual ENSO oscillation, this longer-term oscillation arose from the interactions between the atmosphere and ocean, with the sluggish AMOC setting the considerably slower pace. In 1995, I coauthored an article with Jeffrey Park of Yale and Ray Bradley of UMass Amherst that used a tool that we had developed to identify oscillations in climate data. We applied it to a set of climate proxy records spanning the past 600 years. Our analyses supported the idea of a low-frequency oscillation.[37]

Another article I coauthored in 2000 with Tom Delworth, a climate scientist at the Princeton Geophysical Fluid Dynamics Laboratory, examined climate model simulations, observational data, and long-term proxy reconstructions to argue for the existence of an internal fifty-to-seventy-year oscillation in the climate system, involving the AMOC and the coupling of the ocean and atmosphere in the North Atlantic. Because it was a multidecadal oscillation centered in the North Atlantic, I called it the Atlantic Multidecadal Oscillation (AMO).[38]

The notion that a natural, internal oscillation with a multidecadal (fifty-to-seventy-year) timescale, rather than climate change, might be responsible for an array of climate trends, including tropical Atlantic warming and increases in Atlantic hurricane activity, has since become widespread. I feel like I helped create a monster. Research over the past decade analyzing observations and climate model simulations calls the earlier work into question. As a scientist, you follow where the evidence leads you, and if that puts you in the awkward position of altering your views, so be it. That's how science works. In this case, the evidence has led me to the conclusion that the very phenomenon I helped name, the AMO, doesn't actually exist.[39]

In recent work, my collaborators and I analyzed the state-of-the-art climate model simulations used by the latest IPCC report. We examined the *control* simulations where nothing is changed in the simulations—no volcanic eruptions, solar fluctuations, or human greenhouse gas increases. The model is just left to run on its own, generating internal variability due to weather and its interaction with ocean currents and other components of the climate system. These control simulations do not produce any consistent evidence for internal AMO-like oscillations. In fact, they show no evidence of oscillatory signals other than the well-established interannual ENSO phenomenon. The remaining internal variability is indistinguishable from simple climate "noise."[40]

The apparent AMO signal in the instrumental temperature record is reproduced only in the historical climate model simulations, where the models are driven by human factors, including the long-term increase in greenhouse gases as well as the ramp-up of sulfate aerosols and subsequent ramp-down in the 1970s as a result of the Clean Air Act. What appears to be an "oscillation" is instead seen to be an artifact of the competition between long-term greenhouse warming and the more recent decrease in sulfate aerosol cooling in the late twentieth century, a slow-down and then speed-up of warming that masquerades as a long-term apparent cycle sitting atop a more steady warming trend.

But how does this explain the apparent AMO signal detected in long-term climate proxy data, like those we analyzed back in 1995? In a follow-up publication in *Science* analyzing long-term model simulations of the past millennium, we showed that the apparent multidecadal "cycle" in this case is instead an artifact of natural drivers prior to the historical era. More specifically, it is a consequence of the coincidental pacing that occurred over decades of major explosive volcanic eruptions in past centuries. Take away the volcanic eruptions, and there is no multidecadal "oscillation." Other researchers have independently reached the same conclusion.[41]

So, scientific evidence does not support the notion of an internal, multidecadal, AMO-like climate oscillation. Nor does it support the claim that said oscillation (rather than human-caused warming) is responsible for increases in tropical Atlantic sea surface temperatures and Atlantic hurricane activity. Yet old journalistic habits die hard. A Google news search I performed while writing this paragraph revealed five media reports over the past two weeks attributing increased Atlantic hurricane activity to the AMO.[42]

Climate Sensitivity Revisited. Again.

From the greenhouse climates of the Permian-Triassic extinction and PETM to the icehouse climates of the Pliocene and Pleistocene, we've witnessed an array of examples of how paleoclimate can inform assessments of the crucial quantity known as climate sensitivity.

Although defined in the context of greenhouse warming, ECS can actually be measured from the response of the climate to other drivers, including natural solar and volcanic drivers. The historical climate record itself provides a relatively poor constraint on ECS owing to the shortness of the record and the fact that there are multiple competing drivers, some of which (for example, sulfate aerosol effects) are quite uncertain. Based on this line of evidence alone, there would be a one-in-three chance that ECS is either lower than 1.5°C (2.7°F) or higher than 6°C (11°F). Fortunately, as we know, there are other lines of evidence that inform our assessment of climate sensitivity, helping to reduce the uncertainty range.

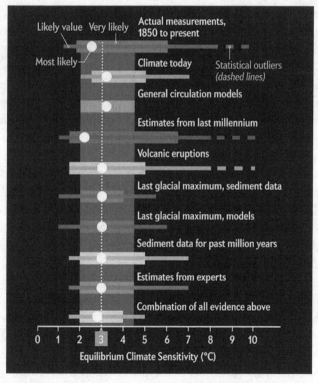

Figure 20. Estimates of ECS from various lines of evidence.

These include, among others, the response of the climate to volcanic eruptions, the cooling during the Last Glacial Maximum, geological evidence of changes in both CO_2 and temperature over millions of years, so-called expert judgment (where you literally poll a group of leading scientists on what they think), and last but not least, comparisons of paleoclimate observations and model simulations over the preindustrial Common Era, typically restricted to the past millennium during which the data are most reliable.

The past millennium studies, curiously, have yielded the lowest apparent values of ECS of all. A prominent 2006 article estimated ECS at around 3.6°F (2°C), a full 2°F (1°C) below most of the other estimates. The study generated the headline "Climate Change Will Be Significant *but Not Extreme*" (emphasis mine) in *The Washington Post*. It would be very good news if it were true. It almost certainly is not, however.[43]

The study employed a simple climate model of the sort described in Chapter 2 where only the global average temperature is calculated. In such a model, ECS is a simple parameter that can be varied. The model is driven by the natural drivers that dominate the preindustrial era and the ECS can be varied until the model simulation past matches the proxy temperature reconstructions. That best match yields an estimate of ECS. The problem is that the estimate is only as good as the data that go into it.

The dominant driver of climate change during the preindustrial Common Era was the cooling that occurred after explosive volcanic eruptions. But there's a problem. There is a mismatch between the sharp volcanic cooling spikes predicted by the models and the more muted cooling seen in the proxy temperature reconstructions. In fact, the 2006 study mentioned above only went back to the late thirteenth century to avoid the largest volcanic eruption of the past millennium, the 1258 CE eruption. The radiative impact of the eruption, based on ice core volcanic aerosol deposits, is estimated to have been four times larger than the massive 1991 Pinatubo eruption. It should have yielded a cooling of 2°C (3.6°F) or more. But there is little or no apparent response at all in the proxy reconstructions of global temperatures. Although this problem is highlighted by the 1258 CE discrepancy, it is more pervasive.

My colleagues and I have argued that this problem is a consequence of the overreliance in these reconstructions on tree cores from trees that grow in very marginal environments of the boreal or alpine tree line where growth is temperature-limited. Such locations are selected because tree growth under these conditions is more likely to be reflective of temperature. However, because these are such cold locales, a large volcanic eruption can cool summers below the minimum threshold for tree growth. Using simulated temperatures over the past millennium and a simple model of tree growth responses, we showed that this would lead both to a loss of sensitivity of tree ring growth (ring thickness) to cooling following large eruptions and chronological errors that accumulate back in time, as growth rings would be missing across large spatial regions in years of zero growth. This effect leads to an attenuation and smearing of the response to very large eruptions that increase back in time—an effect that is observed in tree ring–based summer temperature reconstructions and reproduced by our tree growth model. We estimated that this effect would yield an ECS of around 2.0°C (3.6°F) when the true value is 3.0°C (5.4°F).[44]

Though tree ring researchers have challenged these conclusions, there are additional lines of evidence suggesting that the basic findings are correct. A realignment of tree ring series within the estimated chronological errors yields substantially greater cooling responses that are consistent with model simulations. A separate analysis comparing the reconstructions with climate model simulations of the last millennium led by Andrew Schurer of the University of Edinburgh found that simply removing the few largest volcanic eruptions (which would be most prone to the tree ring underestimation problem) yields inferences consistent with ECS values of around 3°C (5.4°F). Without these additional cross-checks and corrections, one would wrongly conclude—as did the authors of the 2006 study—that climate sensitivity is substantially lower than it likely is.[45]

Here we have yet another reminder that, though the paleoclimate record can provide us with important climate change insights, we have to view that record with a critical eye, ever-cognizant of its limitations and possible sources of bias. In this case, a flawed estimate of climate sensitivity from early last millennium studies likely conveyed a false sense of complacency regarding the prospects for future human-caused warming.

There is actually an even deeper potential problem here that goes beyond the issue of how well proxy data record past climate change. The climate responses during the Common Era are dominated by drivers of *cooling*, like explosive volcanic eruptions. In other words, we're really looking at a "cool" climate response of the climate system. This is important to recognize because ECS is *not* a universal quantity. The relevant feedback processes aren't necessarily the same for cold and warm global climates. In cold climates, like the LGM, ice-related feedbacks are critical, as we saw in Chapter 6. In warm climates, however, there are key carbon cycle feedbacks related to permafrost melt and methane release, or cloud responses that are only triggered in hothouse climates, such as we saw in Chapter 5.

This asymmetry means that ECS values obtained from past cold climates, like the LGM, or cool climates, like the preindustrial Common Era, may not apply to projected future greenhouse warming. A worst-case emissions scenario could take us to CO_2 levels not seen in tens of millions of years. So, we face a catch-22. Our most reliable paleoclimate constraints include the more recent past (for example, the Common Era and the LGM), where both the paleo data and relevant drivers are best known. Yet, both constitute cold or cool climates that are unlikely to exhibit hothouse climate feedbacks and therefore unlikely to reflect hothouse climate sensitivity. We could potentially reach 1200 ppm CO_2 equivalent (that's the equivalent amount of CO_2 we've added to the atmosphere when other human-generated greenhouse gasses like methane and nitrous oxide are included) by the end of the century in a policy scenario of little or no mitigation. We've got to go back to the early Eocene, about fifty million years ago, to find levels that high in the geological past. As we learned in Chapter 5, there are potential new hothouse climate feedbacks that could set in at those levels, raising climate sensitivity.

A 2020 review led by Steve Sherwood of the University of New South Wales, whose work on climate change–induced heat stress we encountered in Chapter 5, combined various lines of paleoclimate evidence in an effort to reduce the current uncertainty range in ECS. The study produced an updated "likely" range of 2.3–4.5°C

(4.1–8.1°F) reduced relative to the canonical 1.5–4.5°C (2.7–8.1°F) range we've encountered before. The reduction in the lower end of the range seems justified—almost no lines of evidence (including the last millennium, when appropriate corrections are made, as discussed above) support values as low as 2.0°C (3.6°F). But is the same upper end of the range still warranted? As we saw in Chapter 5, there is evidence for ECS values as high as 5°C (9°F) during some past hothouse intervals such as the peak of the PETM. But the greatest weight at the upper end of the range in the Sherwood analysis was given to paleoclimate evidence from cold climates, rather than the hothouse climates of the distant past, which leaves me a bit skeptical. Neither the cooling during the largest volcanic eruptions of the Common Era nor the deep freeze of the LGM can tell us anything about feedback processes specific to hothouse climates, the very sorts of hothouse climates we could be headed for in a scenario of climate inaction.[46]

Dangerous Human Interference

In December 2015, at the twenty-first Conference of the Parties (COP21) of the United Nations Framework Convention on Climate Change (UNFCC) in Paris, France, 195 nations adopted what is known colloquially as the Paris Agreement. The agreement commits the countries of the world (all of which have now signed on) to "holding the increase in the global average temperature to well below 2°C [3.6°F] above pre-industrial levels and pursuing efforts to limit the temperature increase to 1.5°C [2.7°F] above pre-industrial levels." Beyond such levels of planetary warming, the impacts of climate change—measured in terms of damaging and deadly weather disasters, coastal inundation, adverse health and mortality, degraded ecosystems, destroyed forests, and threatened oceans—can reasonably be characterized as increasingly catastrophic.[47]

One important source of uncertainty in assessing the carbon emissions budget left for avoiding these danger thresholds involves the determination of just how much warming has already taken place. This might seem an odd source of incertitude, as we have accurate thermometer records that tell us how Earth's average temperature has changed over the past century. But human-caused warming began earlier than that. Leaving aside William Ruddiman's early Anthropocene hypothesis, it is reasonable to define the baseline with respect to which human-caused warming is measured as the beginning of the industrial revolution when large-scale burning of fossil fuels began. That takes us back to the mid-eighteenth century. The convention adopted by the IPCC and many researchers, however, is to simply take the late nineteenth century (for example, 1850 to 1900) as the baseline because that's as far back as reliable global instrumental surface temperature measurements go.[48]

The problem with this convention is that models predict that some human-caused greenhouse warming had already occurred by then. That means that the use of a late nineteenth-century baseline underestimates the warming that has taken place and how close we are to potentially dangerous warming thresholds and tipping points. Because the instrumental records already indicate 1.2°C (2.2°F) of warming since the late nineteenth century, even a tenth of a degree or two has a large impact on how close we are to the 1.5°C (2.7°F) and 2°C (3.6°F) thresholds and the carbon budgets left for avoiding them.

Proxy reconstructions of global mean temperature during the Common Era have uncertainties on the order of several tenths of a degree, limiting their ability to estimate the pre-instrumental warming to within a tenth of a degree or two. Climate model simulations, however, can yield more precise estimates. Andrew Schurer and collaborators (of which I was one) used a set of state-of-the-art last millennium simulations to estimate that around 0.1–0.2°C (0.2–0.4°F) of human-caused warming had taken place prior to the late nineteenth century. Taking into account this additional pre-instrumental warming, Schurer et al. estimated as much as a forty percent reduction in the carbon budget available for avoiding the key target of 2°C (3.6°F) of warming and an even greater reduction in the carbon budget for avoiding 1.5°C (2.7°F). Along with other sources of uncertainty, including how surface air and sea surface temperatures are blended in calculating global mean temperature and how instrumental and model temperature series are merged, the budgets could possibly be even smaller. If we're looking for examples of how uncertainty is not our friend, this is yet another one.[49]

Lessons Learned

The Common Era offers a number of important lessons about the climate crisis. First of all, it alerts us to just how unprecedented the changes taking place today truly are. That stark reality was conveyed by the hockey stick graph more than two decades ago. But the lessons go well beyond the hockey stick.

We see that human-caused greenhouse warming is disrupting many of the subsystems of our climate and altering key modes of climate variability, including the El Niño/Southern Oscillation (ENSO) phenomenon, the Asian summer monsoon, and the great North Atlantic ocean conveyor. Though the science behind these linkages can be complicated, the implications aren't: worsened drought in western North America and other regions around the world, worse flooding in places like South Asia, more active Atlantic hurricane seasons, collapsing fish populations in the North Atlantic, and accelerated sea level rise along the U.S. East Coast.

Paleoclimate data from the Common Era underscore the dramatic increase underway in coastal risk from the twin threats of sea level rise and tropical storm intensification. And they inform important climate policy assessments such as the carbon budget that remains for keeping warming below critical 1.5°C (2.7°F) and 2.0°C (3.6°F) planetary danger limits.

The preindustrial Common Era also affords us an expanded view of natural climate variability. Analyses of the past millennium, for example, cast doubt on the existence of AMO-like internal multidecadal oscillations that have been invoked to argue against the role climate change has played in the observed increase in Atlantic hurricane activity.

With apologies to Clint Eastwood (and millennials too young to get the reference), a scientific discipline has got to know its limitations. So, though paleoclimate evidence from the Common Era provides a number of critical insights, there are also limitations in what inferences we can confidently draw from the available evidence. We should not brush real discrepancies between models and proxy data under the rug, as there is sometimes a tendency for scientists to do. As we saw earlier, these discrepancies may point to limitations in the underlying proxy data that have the potential to render systematic underestimates of policy-informative quantities such as ECS. Finally, though the data might provide solid constraints on the low end of the ECS spectrum, there is reason to be skeptical about efforts to narrow the high end of the spectrum based primarily on cool-climate paleo-climate information such as that provided by the preindustrial past millennium. Observations from hothouse climates suggest the possibility of higher climate sensitivities than inferred from cold climates. We would be unwise to disregard what they tell us.[50]

8

Past Is Prologue. Or Is It?

As for the future, your task is not to foresee it, but to
enable it.

—ANTOINE DE SAINT EXUPÉRY, *Citadelle*

M y late friend and mentor, the great climate scientist and communicator Stephen Schneider, frequently spoke in aphorisms. In
characterizing the climate threat, he once opined that "the end of the
world" and "good for you" are the two "lowest probability outcomes."
The truth, in other words, is almost certainly between those two extremes. As Schneider was also fond of saying, "the truth is bad enough."[1]

OUR EXAMINATION OF KEY climate episodes spanning Earth's history
supports Schneider's pithy characterization of the climate crisis. There
is no need to exaggerate the threat. The facts alone justify immediate
and dramatic action. An objective review of the paleoclimate record
tells us that it's not too late to preempt a truly catastrophic climate
future. The obstacles to action aren't physical or even technological.
At least at this point, they remain entirely political. Employing an
aphorism of my own, there is *urgency*, but there is *agency*, too. The
impacts of climate change, no doubt, constitute an existential threat
if we fail to act. But we *can* act. Our fragile moment can still be
preserved.

Looking Backward

So, what precisely have we learned from our forays into the past? What new perspectives did our journey through the eons impart upon us as regards today's climate crisis? We started from the beginning, the faint glow of our young Sun 4.5 billion years ago, and worked our way toward the hot burning coals of industrialization.

Our assessment of the Faint Young Sun paradox and the Gaia hypothesis underscored the importance of stabilizing feedbacks within the climate system that tend to moderate Earth's climate. We've seen that the planet, and life, are resilient to a great extent. This is observed in the response to slow drivers of change, such as the gradual brightening of the Sun over the eons, the movement of the continents, and the outgassing of carbon dioxide from the solid Earth to the atmosphere.

But we've also seen that shocks to the system can initiate a chain of events that spiral out of control, giving us vicious cycles in place of stabilizing feedbacks. Consider the Paleoproterozoic era more than two billion years ago when the biological innovation of oxygen-generating photosynthesis led to a rapid drawdown of atmospheric carbon dioxide and positive, amplifying feedbacks involving cooling and ice buildup that soon turned the planet into a snowball. The resilience of which we speak evidently has its limits. You can push the system only so far.

We examined the greatest known extinction event in geological history, the Permian-Triassic (P-T) extinction event 250 million years ago when ninety percent of all species on Earth perished. Some point to this event as a simple case of climate-driven extinction, purportedly amplified by methane feedbacks. Now, climate warming certainly played a key role in the extinction event, and seabed methane was a factor. But the real driver was a substantial release of carbon in the form of outgassed CO_2 from Siberian Trap volcanic eruptions. The resulting warming, moreover, was well within the range of what would be expected given conventional estimates of climate sensitivity. That is to say, no "methane bomb" and no massive "hothouse feedbacks," despite the headlines we too often read warning us that such catastrophes already await us.

Moreover, there were a number of factors other than warming that contributed to the mass extinction. Among these were deoxygenation of the atmosphere and ocean, and a deadly hydrogen sulfide "stink bomb" that would have also triggered ozone layer destruction. On top of that, there was ocean acidification both from the volcanic sulfur emissions and the buildup in atmospheric CO_2. Some of these factors are relevant today, but several aren't. Among other things, we lack a single massive continent like the one—Pangea—that existed back then. Pangea easily dried and deforested as the planet warmed, triggering a cascade of impacts, including a decrease in carbon burial, a lowering of atmospheric oxygen, ocean deoxygenation, and ocean hydrogen sulfide poisoning. It is extremely unlikely that human-caused warming would trigger this particular toxic stew of conditions behind the Great Dying. That is not to say, however, that many of these factors—warming and ocean acidification, in particular, due to human carbon emissions—aren't a threat today to us and other living things. They are, and if we fail to rein in our profligate burning of fossil fuels, our own shop of horrors awaits.

The dinosaur-killing K-Pg impact event sixty-six million years ago offers other lessons. The dinosaurs couldn't have foreseen the asteroid strike and were powerless to do anything about it anyway. By contrast, we do see the metaphorical asteroids coming our way—and there's something we *can* do about them. One of them was nuclear winter, a threat tied, rather ironically, to the Cold War. The recognition of that threat was motivated by the discovery of the impact event behind the K-Pg extinction, illustrating the tie-in between bolides and bombs.

But the metaphor is especially apt when it comes to the climate crisis. So apt, in fact, that it was used as a vehicle for the climate crisis–themed satire *Don't Look Up* by film director Adam McKay. Premised on the notion that a massive comet is coming our way and planetary destruction is imminent, the film presented an only thinly veiled allegory for current-day climate inaction. The senior scientist who struggles to inform the public of the imminent threat in the face of intransigent politicians and an indifferent mass media is played by Leonardo DiCaprio. To the great amusement of my friends and family, DiCaprio name-checked me when speaking of his inspiration for the role. Though I will say, given some of the personal foibles exhibited by his character (an episode of marital infidelity among them), I wasn't quite sure this was a good thing! In any case, the point here is that we see the veritable asteroid (or comet) coming our way, and we can still do something about it. There is urgency and there is agency. The threat is existential, but what happens is, still, mostly in our hands.[2]

Our world today is obviously very different from the ancient world of the Faint Young Sun and Snowball Earth. The P-T extinction, as we've seen, is an imperfect analog for today. And there's no evidence that an *actual* planet-smashing asteroid or comet looms in our foreseeable future. So our journey continues. We look to other climate scenarios that might more closely resemble the predicament we face today for additional insight.

That led us to consider the PETM, when initial warming from a series of usually carbon-enriched volcanic eruptions triggered a rapid episode of 7–11°F warming of the planet in as little as 10,000 years. Though the rate of warming still paled in comparison with today's, the PETM is perhaps our best example of a rapid global warming event driven by a massive—albeit natural—release of carbon. Some have speculated that there was enough deep ocean warming to destabilize a large reservoir of seabed methane, adding substantially to the warming. Indeed, the PETM is the canonical example offered up by pessimistic prognosticators of our inevitable and imminent climate doom. They insist it's an analog for the methane-driven runaway warming—and extinction—we purportedly now face.

But the evidence doesn't actually support that narrative. The best available science, including the tools of isotopic analysis, reveals that there was no major methane belch from seabed hydrates during the PETM. The warming, instead, appears to have been largely driven by a rapid initial pulse of CO_2 from volcanic outgassing, followed by a continued slower release over several tens of thousands of years. While estimates of the changes in both CO_2 and temperature are uncertain that far back, they imply a climate sensitivity that falls broadly within the conventional range that we've encountered, if at the slightly higher end of that range. The inferred equilibrium climate sensitivity (ECS) values of 3.7–4.5°C (6.7–8.1°F) are modestly higher than the estimated current-day value of 3°C (5.4°F). Some research, as we saw, suggests that ECS might have even increased *during* the PETM, perhaps approaching 5°C (9°F). Possible hothouse feedbacks—cloud related, for example—might have kicked in as we approached a steamy global temperature of about 90°F (32°C). Even then, though, there was no "runaway warming."

What lessons should we take away from the PETM, then? First, there's the good news: Even when the planet was hotter than a worst-case fossil fuel emission scenario can plausibly make it, there was no runaway warming. There wasn't even a mass extinction—though there were certainly winners and losers as rapid climate change created both challenges and opportunities. Ironically, we—or rather, our ancient primate ancestors—were among the winners. Now, the bad news: Even if PETM-level warmth is out of reach, a policy of total climate inaction could warm the planet up to the point where substantial regions would become uninhabitably hot for human beings—a hotter, more crowded planet with less food and drinkable water. It doesn't take a Venusian runaway greenhouse to yield a dystopian future. We would be the losers in that scenario.

The PETM, in any event, is still an imperfect analog because the starting point for the warming—the balmy greenhouse of the early Eocene—was hot compared to today. A better analog, arguably, can be found on a slightly icier world more like today. That led us into the late Cenozoic Ige Age where the ice sheets that are so familiar to us today—the Greenland (GIS) and Antarctic (AIS) Ice Sheets—first came into being. The mid-Pliocene, around three million years ago, might seem like an especially appropriate analog for today, with a CO_2 level of about 400 ppm that rivals the present level.

At first blush, the evidence from the mid-Pliocene seems ominous. At CO_2 levels similar to or slightly lower than today, global temperatures were at least 2°F warmer than today. Sea level by some accounts was more than eighty feet higher than it is now. That implies no GIS, no West Antarctic Ice Sheet (WAIS), and a chunk of the East Antarctic Ice Sheet (EAIS) missing as well. Such evidence is sometimes cited to imply that current CO_2 levels, if they are simply maintained where they are, will give us an additional 2–3.5°F warming and the better part of a hundred feet of sea level rise. According to sundry, breathless press releases and media accounts, we must begin adapting to a mid-Pliocene world that would only be familiar to Noah and his ark.[3]

But that isn't true, either. First of all, the oft-cited eighty feet of sea level rise, as we learned in Chapter 6, is probably the product of misinterpreted sedimentary evidence that fails to properly account for geological effects such as isostatic rebound. The actual sea level rise might have been more like thirty feet relative to today. Still bad, for certain. But not *Waterworld*.

More significantly, the comparisons to the mid-Pliocene ignore the physical phenomenon of hysteresis with which we are now familiar. Just because the GIS didn't exist when CO_2 levels had been slowly lowered to 400 ppm over the course of the Cenozoic CO_2 drawdown doesn't mean that it is doomed upon reaching similar levels from the human-caused CO_2 increase. Once you have the ice sheet, as we do today, it is resilient—to a point. And with the GIS in place, global temperatures are cooler than they would otherwise be.

So, no, we're not committed to a mid-Pliocene world yet. Lest we be too sanguine about the current state of affairs, however, suffice it to say that hysteresis only buys you so much of a cushion. It is conceivable that just another 1°F or so of warming might be enough to push the GIS over the edge. The fact that we're already seeing substantial surface melt from Greenland is warning enough.

Now leap forward into the Pleistocene, past the first eight 100,000-year sawtooth cycles, all the way to the penultimate interglacial period, the Eemian. The levels of CO_2 then were comparable to the modern preindustrial level (about 280 ppm), and global temperatures were similar to today. The geological evidence indicates that global sea level was at least twenty feet higher than today, which requires substantial melt contributions from Antarctica and some of Greenland. Assessing the implications for us today is complicated, however, because both CO_2 and temperatures were coupled during the Pleistocene glacial cycles in a way that makes it challenging to isolate a simple causal relationship between CO_2 and warming.[4]

The unusual Earth orbital configuration was especially conducive to warm high-latitude summers, favoring the observed Eemian ice sheet recession. Though global temperatures were comparable to today, Arctic summers at least 3.5°F warmer than now were primarily responsible for Greenland ice loss. That's good news for us today, because we can still avoid that warming. Under aggressive policies where carbon emissions are ramped down to zero in the decades ahead, Arctic summers are almost certain to remain less than 3.5°F warmer than today. Even under current policies alone, without further action, they are likely to. Alternatively, in a worst-case scenario of "no policy," where we roll back existing restrictions and commitments, we will likely breach the Arctic summer 3.5°F warming threshold by mid-century. The future, once again, is in our hands.[5]

As we know, both polar bears and Arctic sea ice survived the Eemian, and there's no evidence that there was a massive release of permafrost methane despite the outsized Arctic warming that took place. The Eemian thus provides us with one of the more compelling arguments against the premise, popular among climate doomists, that Arctic "methane bomb" warming is either underway or imminent. The truth, it bears repeating, is bad enough.[6]

The flip side of the Eemian was the Last Glacial Maximum (LGM), when orbital factors instead favored cold high-latitude summers and peak continental glaciation. Estimates of ECS diagnosed carefully from the cooling of the LGM—which account for not just the lower CO_2 levels but the various other relevant drivers, including dust and the increased albedo of more widespread continental ice cover—point to ECS values of about 3°C (5.4°F). Simulations that reproduce not only the LGM but the full history of the mid- to late Pleistocene 100,000-year cycles, point to similar ECS values.

While we might take some solace from these numbers, which suggest moderate levels of climate sensitivity, we're reminded that constraints derived from the icehouse climate of the Pleistocene may not adequately reflect the feedbacks and processes that characterize hothouse climates like the ones we could hypothetically be headed toward.

Today, scientific differences of opinion are sometimes litigated in real time on social media. Back in October 2020, I was involved in an online exchange involving various scientists whose work is now familiar to the reader: Gavin Schmidt, a coauthor of the Sherwood et al. article discussed in Chapter 7; Matt Huber, whose hothouse climate sensitivity work we encountered in Chapter 5; and University of Arizona paleoclimatologist Jessica Tierney, who led the LGM climate sensitivity article discussed in Chapter 6. The discussion involved this very matter: whether or not we can constrain the upper end of the climate sensitivity uncertainty range based primarily on information from the icehouse climates of the Pleistocene. Schmidt and Tierney weighed in on the affirmative, while I sided with Huber weighing in on the negative. There are honest differences of opinion among experts on such matters. In my view, though, the greenhouse remains murky. I worry about surprises that lurk therein.[7]

Finally, we come to the Common Era, the interval spanning the past two millennia within which we currently reside. The lessons here are about as subtle as a hockey stick to the behind. We are engaged in an unprecedented experiment with the planet. The profundity of this dangerous experiment is betrayed by graphs depicting the unprecedented warming of the past century and the rise in sea level that has accompanied it. Measured relative to the background of the muted trends in centuries past, the impact of fossil fuel burning and other carbon-generating human activities of the industrial era is at once obvious and stark. The fragility of our current moment is laid bare.

With 4.5 billion years' worth of paleoclimate lessons now at our disposal, let us instead look forward, using the fuzzy crystal ball that climate science affords us. We cannot say what our future will be. But we can talk about what futures we are still able to create.

Looking Forward

It is important to recognize that there are limitations to what paleoclimate studies can tell us when it comes to the outstanding scientific questions that remain about the climate crisis. We've seen many of them: the potential for proxy data to underestimate some changes, and the difficulty in untangling the multiple factors, climatic and otherwise, that may impact what is recorded by a particular proxy record. The past is somewhat murky, more so the further back in time we go, as less data are available and those that are available suffer from possible alteration. There is no shame in acknowledging that paleoclimate studies cannot address all outstanding questions regarding climate dynamics, climate variability, and climate change. The evidence from the paleoclimate record should instead be viewed as one very valuable source of information that, combined with other sources, can yield a fuller understanding of the climate system and the climate crisis.

In the fictional climate-themed disaster film *The Day After Tomorrow*, the protagonist—a paleoclimatologist played by Dennis Quaid—explains to a desperate government operative that all he has to offer is "a reconstruction of a prehistoric climate shift. It's not a forecast model." He nonetheless goes ahead and uses it as a forecast model, predicting, with stunning precision of course, the climate catastrophe that plays out in the film. But that's not the real world. And Dennis Quaid was right the first time. The most fundamental limitation of paleoclimate inferences is that, though they may inform our understanding of the climate system, they cannot be used to make an actual prediction about the future. Past is *not* always prologue. So, we turn to prognostic models that apply the laws of physics to the Earth system to make those future predictions.

The IPCC has adopted various scenarios called Representative Concentration Pathways (RCPs) that describe different possible carbon emissions pathways that make different assumptions about future climate policy. They are not predictions, but they are useful guidelines for how different policy choices lead to different levels of future planetary warming. Scenario RCP2.6 represents what we might call substantial climate action, keeping equivalent CO_2 concentrations below 450 ppm and limiting warming (since the preindustrial era) to about 2°C (3.6°F). RCP4.5 represents what we might call current policies, keeping CO_2 concentrations below 560 ppm (twice preindustrial levels) and limiting warming below about 3°C (5.4°F). RCP8.5 might reasonably be called a worst-case scenario. The levels of CO_2 exceed 1200 ppm—more than four times preindustrial levels (two doublings)—by the end of the century. Such a scenario is consistent with either a policy of total inaction (including a reversal of climate policies already in place) or a policy of weak action coupled with greater-than-estimated amplifying carbon cycle feedbacks resulting from forest burning or methane release. Global temperatures in this scenario would likely exceed about 5°C (9°F) relative to preindustrial levels by the end of the century and about 9°C (16°F) by 2300. Indeed, the high end of the uncertainty range among models indicates the possibility of as much as 13°C (23°F) warming.

The paleoclimate record places these model projections in a stark context. In the worst-case scenario, we would leave the icehouse conditions of the late Cenozoic behind, returning to the greenhouse conditions of the Eocene. By the end of this century, global temperatures would rival those of the late Eocene, about thirty-five million years ago, and by 2300 they would rival the mid-Eocene, about forty-five million years ago. If we are unlucky, and the very upper end range of the model's spread proves correct, the warming of 13°C (23°F) would give us an average Earth temperature of 28°C (82°F), rivaling the early Eocene.

As we learned in our discussion of the steamy Eocene in Chapter 5, a world this hot would be mostly uninhabitable to human beings, with wet bulb temperatures in the dangerous-to-deadly range of upper 80s°F encountered regularly over a large region of the planet. That's a world we can easily avoid with reasonable efforts to mitigate carbon emissions. But deadly heat cannot be avoided altogether. Even in a scenario of substantial climate action where warming is kept below 2°C (3.6°F)—or, for that matter, a scenario of dramatic and immediate action, which is necessary to keep warming below 1.5°C (2.7°F)—we will still witness elevated risk of dangerous heat exposure. That's an easy prediction to make because we're already seeing evidence of it today.

The 2003 European heat wave that saw 30,000 people perish was a sign of things to come. Recent studies suggest that we are now experiencing as many as five million deaths a year from heat stress and other dangerous weather extremes. Add in the additional four million deaths a year from air pollution generated by fossil fuel burning, and that's nine million. In one year, the combined deaths are nearly twice the total recorded number of deaths worldwide from the COVID-19 pandemic. Fossil fuel burning and the climate change it causes is far more deadly than the worst pandemic the world has yet faced.[8]

Adverse health impacts of heat stress, as we learned earlier, are already being felt in various cities in the western United States. One of those cities is Phoenix, Arizona. Some readers may recall a news

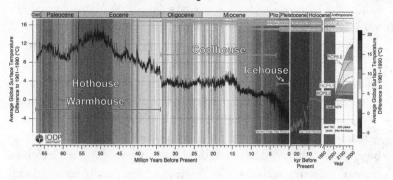

Figure 21. Model-based projections of future warming under different carbon emissions scenarios in the context of the paleoclimate record of the Cenozoic.

story from June 2017. Phoenix, already known for its withering summer heat, was experiencing a record June heat wave. On June 20, 2017, temperatures at the Phoenix airport rose to 120°F, exceeding the maximum safe operating temperature (118°F) for aircraft takeoff. Warm air is less dense than cold air. Once the air near the surface reaches 118°F, it no longer provides enough buoyancy for liftoff to be achieved before the accelerating aircraft reaches the end of the runway. So all flights were grounded.[9]

I experienced this event myself. I had just returned with my family from a vacation to the Grand Canyon, and we were staying overnight at a hotel near the Phoenix airport before flying home the next morning. My daughter and I went swimming in the outdoor pool that afternoon. It was like diving into a bathtub. Treading barefoot on the pavement was like walking on hot coals. Needless to say, it didn't take long for us to retreat to the oasis of our air-conditioned hotel room.

That night my wife and I were awakened around 3:00 a.m. by our then eleven-year-old daughter. She was in distress, having trouble breathing. We hired a taxi and ushered her to the emergency clinic of the nearest hospital. The doctor put her on an inhaler, and that seemed to alleviate her symptoms for the time being. We took her to see the doctor once we were back home, and he diagnosed her as suffering from intermittent asthma, which was likely triggered by the dangerously high surface ozone levels resulting from the record heat.

That was the moment when I first came face-to-face with the danger-ous impacts of climate change. That's when it hit home—literally.[10]

I personally faced the climate crisis again during the austral sum-mer of 2019/2020, which has come to be known as the Black Sum-mer. I had arrived in Sydney for a sabbatical, researching the impact of climate change on extreme weather events in Australia. Instead, I witnessed those impacts play out in real time in the form of a com-bustible mix of record heat and drought that turned Australia into an infernal hellscape with massive, destructive, and deadly bushfires that spread out across the continent.[11]

Of course, similar episodes have been experienced by millions of people around the world now. In Chapter 5, we talked about the months-long heat wave that impacted India and Pakistan in spring 2022. With wet bulb temperatures that exceeded 86°F, dozens of peo-ple succumbed to heat stroke. According to IPCC lead author Dr. Chandni Singh, the record heat wave tested "the limits of human sur-vivability." The persistent extreme heat led to power blackouts and adversely impacted wheat yields, underscoring the cascading impacts of these extreme heat events.[12]

The United States has experienced summer after summer of re-cord heat. June 2021 witnessed the infamous "heat dome" event that set temperature records in the Pacific Northwest, with high tempera-tures of 116°F in Portland, 108°F in Seattle, and triple-digit heat ex-tending well into southwestern Canada. A group of scientists behind the World Weather Attribution project used a climate model–based detection and attribution approach to assess the role that climate change played in this event. They estimated that, without accounting for climate change, it was a 150,000-year event. In other words, it shouldn't have happened even once over the course of the entire last glacial/interglacial cycle. But what's even more eye-opening is that they estimated it was *still* a rather implausible *thousand-year* event when *accounting* for the impact of human-caused warming.[13]

Do we dismiss the "heat dome" then simply as bad luck? An unfor-tunate roll of the weather dice? Or do we recognize that models might be failing to capture some important mechanisms that are contribut-ing to these extreme events? I argued the latter in an op-ed in the *New York Times* with my friend and colleague Susan Joy Hassol. We noted, among other things, that the phenomenon of resonance—which is not well captured in current-day climate models—was at play in the jet stream configuration that set up this unprecedented event.[14]

Resonance is a phenomenon that applies to wave-like disturbances, from the quantum all the way up to planetary scale. It is perhaps most familiar to us when we sing in the shower. Our voices are amplified because the physical dimensions of the shower (four feet or so wide) correspond closely to the wavelengths of sound that we produce. The waves are trapped by the walls of the shower and as a result grow in amplitude. A similar phenomenon can occur with the atmospheric wave disturbances called Rossby waves or planetary waves. They are familiar to us as the north-south wiggles, meanders, and undulations in the jet stream we observe on weather maps.

My colleagues and I have shown that this resonance phenomenon is favored by the accelerated warming of the Arctic. The decreased contrast in temperature between the cold Arctic and warm subtropics causes the jet stream to slow down and, under the right circumstances, like the ones that prevailed in early June 2021, settle into a very wiggly and stable configuration. Resonance helped the very deep, high-pressure center, or ridge, set up out west and become locked in place, where it grew into the dangerously hot "heat dome."[15]

In June 2022, we witnessed a remarkable hemisphere-wide array of extreme weather events tied to jet stream resonance. Among them was the record heat in North America with triple-digit heat indices covering large swaths of the central and eastern United States. A third of the American population found itself subject to dangerously hot conditions. Record flooding in Montana meanwhile ravaged hallowed Yellowstone National Park. It was an example of an increasingly common phenomenon—a hybrid extreme weather event. In this case, a massive pulse of glacial meltwater due to an unusually warm spring combined with a stationary, low-pressure system that dumped three months' worth of rainfall in a few days. Another increasingly common type of hybrid event is when extreme summer drought and wildfire, with attendant vegetation destruction and destabilization of topsoil, is followed by deadly mudslides when flooding winter rains subsequently arrive, spiked by a warmer atmosphere with more precipitable moisture. Such a situation was tragically on display in the winter of 2017/2018 when nearly two dozen people lost their lives in Southern California. A similar number of Californians lost their lives during the "bomb cyclones" of December 2022 and January 2023.[16]

During the same June 2022 resonance event that was responsible for extremes of heat and rainfall in North America, more of the same was seen further downstream. Europe, too, was experiencing record heat. In France, all-time heat records fell in many locations in early June, nearly two months before typical peak summer heat. Northern Italy suffered not just from excessive heat but also an ongoing drought where 100 days had passed without a single raindrop. And in India, which just can't seem to catch a break, millions of homes were submerged in record floods. In mid-summer 2018, a similar array of extreme weather events that broke out across North America, Europe, and Asia was connected with resonance.[17]

Why am I belaboring this point about the jet stream, wave resonance, and extreme weather events? It has to do with the issue of scientific uncertainty. Uncertainty is often cited by critics of climate action as if it is justification for inaction or delay. But it's just the opposite. In many cases it's not cutting in our favor but against us. In Chapters 6 and 7, we saw that earlier climate models underestimated the potential for ice sheet collapse and sea level rise due to the absence of key processes related to ice sheet collapse in the models. As those processes began to be included in the models, the projections steadily rose, and now rather than talking about one foot of sea level rise by 2100, we're talking about the very real possibility of six to seven feet.

Something similar is true of extreme weather events. My collaborators and I have shown that current climate models are unable to accurately capture the processes involved in jet stream wave resonance events. This is problematic because, as we have seen now, this mechanism is implicated in many of the damaging extreme weather events that we have witnessed in recent years, and there is evidence that climate change is making resonance events more frequent. Both the diagnosis of the impact that climate change is already having on extreme events and projections of future increases in those events are likely underestimated by current-generation climate models.[18]

Climate models, once again, are fuzzy, rather than clear crystal balls. They provide important guidance, in many respects our best guidance, drawing upon the laws of physics, chemistry, and biology to make quantitative, rigorous projections of our potential futures. That's a whole lot better than relying upon hunches, opinions, and wild speculation. The overall warming of the planet, for example, is very much in line with early climate model predictions. But when it comes to *some* key climate change impacts, such as ice sheet collapse, sea level rise, the retreat of arctic sea ice, ocean conveyor slowdown or collapse, western North American drought, and the increase in extreme weather events, the absence or poor representation of important processes in the models leads to a systematic underestimate of the rate and magnitude of the changes.[19]

Such nuanced views struggle to gain currency in a political economy where hot takes, hyperbole, and polarizing commentary best generate clicks, shares, and retweets. I often encounter, especially on social media, individuals who are convinced that the latest extreme weather event is confirmation that the climate crisis is far worse than we thought, and scientists and climate communicators are intentionally "hiding" the scary truth from the public. It is the sort of conspiratorial thinking that we used to find among climate change deniers, but increasingly today we see it with climate doomists. Such sentiment emerged, for example, during the mid-June 2022 heat wave, where one individual tweeted at me and my climate scientist colleague Katharine Hayhoe: "Again we see that climate science as often presented to the public is too conservative, avoids what at the time are deemed worse [sic] case scenarios. BUT these are becoming our reality TODAY."[20]

This is not true, or at best partly true. I responded, "Actually, the warming of the planet is very much in line with early climate model predictions. Some impacts, such as ice sheet melt and sea level rise, and the slowdown of the ocean 'conveyor belt' are exceeding those predictions." Current policies alone, as we've seen, likely keep warming below 3°C (5.4°F), nowhere near the "worst-case" scenarios that we've discussed. That doesn't mean that some impacts aren't unfolding earlier and more dramatically. They are. Again, as the great Stephen Schneider counseled decades ago, it's neither "end of the world" or "good for you." The collective evidence supports *neither* fatalism *nor* complacency.[21]

It is also important to recognize that climate change isn't a cliff that we go off at certain thresholds of planetary warming such as the oft-discussed 1.5°C (2.7°F) warming level, though it is often framed that way. Climate action isn't a binary case of "success" or "failure." A better analogy is that it's a dangerous highway we're going down. We need to take the earliest exit ramp possible. Dangerous climate change impacts, as we have seen, are already being felt—in the form of devastating droughts, heat waves, wildfires, floods, and super-storms. Supply chains have been disrupted through a combination of a pandemic—which is likely at least in part a result of ecological destruction—and more extreme weather, sometimes with disastrous consequences, such as shortages of baby formula. Extreme heat is leading to substantial decreases in worker productivity, costing the U.S. economy alone nearly 100 billion dollars a year.[22]

Dangerous climate change cannot be avoided. It's already here. So, it's a matter of how bad we're willing to let it get. Worse impacts can be avoided if we limit the warming below 1.5°C (2.7°F). But if we miss that exit off the carbon emissions highway, 2°C (3.6°F) is certainly preferable to 2.5°C (4.5°F). And if we miss *that* exit, 2.5°C (4.5°F) is certainly preferable to 3°C (5.4°F). Consider, for example, the matter of species extinction. The IPCC estimates as much as fourteen percent of species could be lost at 1.5°C (2.7°F) warming and eighteen percent at 2°C (3.6°F). Tragic for sure, but greater rates of extinction are expected from other unchecked human activities, including habitat destruction and human exploitation of animals. However, the number climbs to twenty-nine percent at 3°C (5.4°F), thirty-nine percent at 4°C (7.2°F), and forty-eight percent at 5°C (9°F). Half of all species would, by any reasonable standard, consti-tute a sixth extinction event rivaling the great extinctions of Earth's geological past. But that is avoidable in a scenario of meaningful climate action.[23]

Despite the breathless claims of climate-driven mass extinction that one sees all too often in today's headlines, we are not yet re-motely committed to such a future. We can avoid catastrophic cli-mate impacts *if we take meaningful actions to address the climate crisis*. Yes, that's an important "if." But the science actually tells us it's doable. One of the important developments in climate science over the past decade is the recognition that greenhouse warming depends on cumulative carbon emissions up to a given point in time. This has led to the concept of the carbon budget, which determines how much additional carbon we can afford to burn and still limit warming to below a particular level.

The conventional wisdom was once that surface warming would continue on for decades even if we stopped emitting carbon into the atmosphere due to the sluggishness of the oceans, which continue to warm up even after CO_2 stops increasing. This is known as committed warming. But committed warming is only half of the story, an artifact of simplistic early climate modeling experiments in which CO_2 levels were kept fixed after the hypothetical cessation of emissions. Later, more comprehensive simulations with interactive ocean carbon cycle dynamics revealed that CO_2 levels actually drop after emissions cease as the oceans continue to draw carbon down from the atmosphere. That decrease in the greenhouse effect cancels out the committed warming, and the result is an essentially flat line. In other words, global temperatures stabilize quickly once net carbon emissions drop to zero.[24]

As a consequence, we can calculate the carbon budget for a particular global temperature stabilization target. To keep surface temperatures below 1.5°C (2.7°F), for example, carbon emissions have to be brought to zero within three decades, and we have to get halfway to zero within a decade. There are some confounding factors. For example, when coal burning ends, there is a drop in cooling sulfate aerosol pollution, which leads to warming. But that warming is largely offset by a decrease in other warming factors, including greenhouse gases like methane and black carbon from fossil fuel burning. These additional factors all nearly cancel one another out as well.[25]

There are scenarios where global temperatures exceed a given target such as 1.5°C (2.7°F), rise as high as 2°C (3.6°F) or so by mid-century, and then come back down and stabilize below 1.5°C (2.7°F). This is called overshoot, and a shorter-duration, small overshoot is favorable, from a climate-impact standpoint, to a longer-duration, large overshoot. Once again, there are no absolutes. The less, and shorter duration the warming, the better. But the most comprehensive and authoritative assessment of risk across all sectors—health, food, water, conflict, poverty, and the natural ecosystem—by the IPCC in 2018 basically concluded that we don't want to warm the planet beyond 1.5°C (2.7°F), and we *really* don't want to warm it beyond 2°C (3.6°F). And if we do happen to overshoot those targets, we want to keep the duration of overshoot to a minimum.

Where do we stand in this effort? Scientists have evaluated the upwardly revised commitments made at the United Nations Climate Change Conference (COP26) in Glasgow in late 2021 and have determined that they would likely keep warming below 2°C (3.6°F). That's substantial progress compared with the roughly 4°C (7.2°F) warming that we were headed toward prior to the 2015 Paris summit. But it's still a lot riskier than limiting warming to 1.5°C (2.7°F). Moreover, it's one thing to make commitments, and something else entirely to keep them. As my colleague Susan Joy Hassol and I explained in a *Los Angeles Times* op-ed published at the completion of COP26, the goal of limiting warming to 1.5°C (2.7°F) is still alive but "*only if the hard work begins now*" (emphasis added).[26]

Among other things, a pathway to 1.5°C (2.7°F) requires there be no new fossil fuel infrastructure at a time when pipelines continue to be built. A handful of fossil fuel companies—including Exxon-Mobil and Gazprom (Russian state fossil fuel company)—are planning for new projects that will produce about 200 billion barrels of oil and gas. That's the equivalent of a decade of emissions from China, the world's largest current producer of carbon pollution (the United States, meanwhile, is the world's greatest cumulative source of carbon pollution).[27]

Holding policymakers, opinion leaders, and corporations accountable is essential. For while citizens themselves now overwhelmingly support concerted climate action, they can't effect the needed changes themselves. We, as individuals, can of course make climate-friendly choices as consumers. But we cannot impose subsidies for the renewable energy industry or remove them for the fossil fuel industry, price carbon, or block major fossil fuel infrastructure projects. It is only our elected policymakers who are in a position to do that. In the United States, one of the two major parties, the Republican Party, is largely beholden to the fossil fuel industry. And it has acted that way.[28]

The greatest obstacle to climate action, as I detailed in *The New Climate War*, is a sustained, massive, billions-of-dollars disinformation campaign by the fossil fuel industry. Equally culpable are its abettors in the conservative media. And none is more implicated than Rupert Murdoch, a close long-term business partner of the world's largest oil exporter, the Saudi Arabian royal family. Yet, we have something to learn from our friends in Australia when it comes to fighting back.[29]

Murdoch has wielded News Corp as a weapon against climate action, using his international media empire to promote climate change denialism and fossil fuel industry disinformation. His news outlets include the notorious Fox News, which has single-handedly poisoned the minds of its millions of American viewers with fossil fuel industry propaganda for years. But Murdoch has an even greater stranglehold on the Australian media environment, where he controls nearly two thirds of its total newspaper circulation, as well as major television networks Sky News and Foxtel.[30]

And yet we see cracks emerging here in Murdoch's echo chamber of disinformation. A bit of history is in order. Much as is the case with the United States, where Republicans once supported environmental protection and climate action specifically, there was a time in Australia when both major parties (the progressive Labor party and the conservative party known as the Liberals) agreed on climate policy. A cap-and-trade system to limit carbon emissions, known as the emissions trading scheme (ETS), was first proposed by Liberal prime minister John Howard in 2006, and the first piece of legislation to implement it was introduced by Malcolm Turnbull, then the environment minister for the Liberal party.[31]

Labor leader Kevin Rudd, who had worked with Howard on climate policy, was elected prime minister in the 2007 election and attempted to implement the ETS. But a strange-bedfellows coalition of the opposition Liberals—led by climate change denier Tony Abbott—and the Greens—who didn't think it went far enough—opposed it. In 2011, Rudd's Labor party colleague Julia Gillard replaced him as prime minister and successfully passed the ETS, making Australia the first major industrial country, after the European Union, to implement market mechanisms to limit carbon emissions.

Australian fossil fuel interests quickly mobilized against the ETS. The usual suspects—fossil fuel interests and the Murdoch media—savaged Gillard. They portrayed the ETS, as the *New York Times* put it, "as a burden that would hurt businesses and cost households, instead of one that would cut pollution and ensure a more secure future for our children." The reality is that price increases were minimal, and revenue was returned to consumers, with low-income earners actually benefiting. But it didn't matter, the damage was done. Labor was defeated in the next election. And though Turnbull, who had been leader of the Liberal party, supported the ETS, a growing insurgency on the right of the Liberal and National parties, supported by fossil fuel interests and the Murdoch media, staged a coup that replaced Turnbull with climate change denier and fossil fuel apologist Abbott.[32]

But a movement was brewing among independent politicians aligned with neither the Labor nor the Liberal party and supportive of climate action. They call themselves "teal independents," teal being a combination of green (signifying a prioritization of the environment) and blue (the traditional color associated with the Liberal party). The very first of them was Zali Steggall of Warringah, a wealthy coastal electoral district of Sydney. She obtained her seat by beating the incumbent Liberal MP during the May 2019 Australian election. She made climate change a central focus of the campaign. Her opponent was a climate change denier who opposed action. He also had a history of misogynistic attacks against female political opponents. He had held his seat for twenty-five years until he was defeated by Steggall. His name? Tony Abbott.[33]

In the May 2022 Australian national election, the Murdoch propaganda machine failed to elect a fossil fuel–friendly conservative government. Climate became a defining issue in the election. The enduring legacy of the Black Summer was certainly part of the reason. But we must not understate the role of the rise of the teal independents—and the tectonic shift in Australian politics that it represents. Labor, the Greens—who gained two seats in the lower house—and last, but not least, the teal independents all campaigned on a platform of greater climate action and more ambitious emissions targets. The new Labor party government, led by Anthony Albanese, has pledged a forty-three percent reduction in emissions by 2030.

In an op-ed with Malcolm Turnbull, whom I got to know during my Black Summer sabbatical down under, I discussed the lessons that Americans might draw from what had transpired in Australia. How, for example, did Australia manage to defeat the Murdoch climate disinformation machine, which has so effectively waged war on climate policy in the United States for years?[34]

Several features of Australia's electoral system made it resistant to Murdoch's corrupting influence. The lack of gerrymandering and compulsory voting both contribute to more democratic political representation. Both of these goals are laudable in the United States, but implementing them would be an uphill battle given likely partisan opposition in red states. Finally, and most important, though, is Australia's ranked choice voting. Right now, ranked choice is the law in only two U.S. states. Interestingly though, they're not blue states: one is purple (Maine) and the other is deep red (Alaska). Arguably, this is why the two Republican senators from those states—Lisa Murkowski and Susan Collins—are more centrist in their policies than the vast majority of their caucus.

A growing number of cities and municipalities have implemented ranked choice voting, and twenty-nine states are currently considering implementing it. It enjoys high levels of bipartisan support—something that is quite rare in today's hyperpartisan American political environment. Its potentially game-changing impact was evident in the performance of the teal independents in Australia. They were running against Liberal incumbents, most of whom would normally get a first-preference vote of fifty percent or more. If the teal independents could eat into that vote just ten percent, bringing it down to forty percent or less, and then run second among a large number of voters, they could win with the benefit of support from Labor and Green voters. That's what happened.

As Malcolm Turnbull and I stated in the closing of our commentary, "People power trumped Murdoch. Perhaps it can do so in the US." Given the United States' role as the largest cumulative carbon polluter, American leadership is critical in generating support among other major polluters, including China and India. This is one way change might be achieved. We've been given a road map for the path forward on climate.[35]

Ironically, it is at this very moment of promise that a new obstacle has emerged. The greatest threat is no longer denialism—which is frankly untenable given the impacts that we can all see playing out in real time—but rather doomism, the notion that it's too late to act.

Doomism takes various forms, but one of the most prominent examples is the movement known as Deep Adaptation, founded by Jem Bendell, an academic from the United Kingdom. Back in early 2019, Bendell posted a document that the media outlet *Vice* called "the climate change paper so depressing it's sending people to therapy." But it wasn't a peer-reviewed academic article. In fact, it was rejected by scientific journals, and Bendell ultimately self-published it on his personal website. Despite the distinct lack of academic rigor, the paper has nonetheless been viewed far more than any typical peer-reviewed article, with more than 100,000 people having read it. Although Bendell's doomism is better disguised than, say, that of other prophets of despair such as Guy McPherson who insisted in 2012 that runaway global warming would kill the vast majority of human beings by 2020, Bendell nonetheless argues that "climate-induced societal collapse is now inevitable in the near term," which he says means about a decade (that was five years ago). Bendell bases these claims on the discredited premise of an imminent Arctic "methane bomb" and runaway warming, which he insists will cause the collapse of agriculture, exponential increases in infectious disease, and possibly—he implies—human extinction. Deep Adaptation now has a huge cult following despite its total lack of scientific credibility.[36]

There is certainly a worthy conversation to be had about the threat of societal collapse. At a time when we continue to do damage to our global environment in the form of deforestation, air and water pollution, overfishing, and fossil fuel extraction—at a time when we find ourselves on a collision course with basic planetary boundaries of sustainability, with new pandemics arising out of habitat destruction—there is cause to doubt the viability of a continuing policy of extraction and natural resource–driven growth. In an era where misinformation and disinformation run rampant, weaponized by bad actors feeding on native grievance and resentment to advance an agenda of authoritarianism and fascism, there is ample cause for worry. If you're not the least bit worried, you're not paying attention.

It is understandable that some climate advocates have grown frustrated by the slow and still insufficient policies that have been adopted thus far. But they shouldn't allow that frustration to be seized upon and co-opted by false prophets who would lead them down a road of disengagement and inaction. Yes, we have a massive challenge on our hands. But breathless claims of imminent climate-driven "human extinction" and "runaway warming" are both scientifically unsupportable and unhelpful. As Susan Joy Hassol and I wrote in *Time* magazine, "there is no point beyond which we shouldn't keep trying to limit warming. Every fraction of a degree matters to the level of suffering climate disruption will rain down on us."[37]

Rather than being directed into doomism and depression, frustration needs to be channeled toward justified righteous anger. Research shows that, unlike emotions such as fear, anxiety, and depression, anger actually leads to empowerment, engagement, and action. In short, we must recognize the true enemy—bad actors in the fossil fuel industry and their abettors—and channel that anger and frustration into political action.[38]

Our review of Earth's climate history provides us reason for hope in this venture. It has shown us that neither runaway methane-driven warming nor an unhabitable PETM-like hothouse world are plausible in any scenario but total inaction (combined with very bad luck). That doesn't mean that there isn't reason for concern. A runaway warming scenario is hardly necessary to threaten the stability of human civilization. We already see challenges to that stability from political movements that are underway and growing competition over diminishing natural resources. Climate change adds fuel to the fire, decreasing productivity, interrupting supply chains and distribution systems, fanning the flames of conflict over water, food, and land, and nurturing destabilizing pandemics.

So, we return to the fundamental question that this book has sought to address: Is our climate today precariously perched on a knife edge, ready to collapse in a steaming death spiral of methane-driven runaway warming? Or is it resilient enough to tolerate continued fossil fuel burning with minimal consequences? The answer, as Stephen Schneider counseled decades ago, is *neither*.

Even under a business-as-usual scenario where we fail to build on climate policies already in place, the warming of the planet is unlikely to exceed 3°C (5.4°F). No "methane bombs," no runaway warming, no Hothouse Earth. But at that level of warming, we can expect a lot of suffering, species extinction, loss of life, destabilization of societal infrastructure, chaos, and conflict. An end, perhaps, to our fragile moment.

That's not a world in which we want to live, and it's not a world that we want to leave behind for our children and grandchildren. Though it is a possible future, it's not a preordained future. If we build upon the actions that have already been taken, decarbonize the machinery of our civilization in the years and decades ahead, we can preserve our fragile moment.

That's what our review of climates past and present convincingly tells us. So let that be our rallying call and our mission.

Acknowledgments

I am grateful to the many individuals who have provided help and support over the years. First and foremost are my family: my wife, Lorraine; daughter, Megan; parents, Larry and Paula; brothers, Jay and Jonathan; and the rest of the Manns, Sonsteins, Finesods, and Santys.

I am indebted to all those who have inspired me, mentored me, and served as a role model to follow, including, but not limited to, Carl Sagan, Stephen Schneider, Jane Lubchenko, John Holdren, Bill Nye, Paul Ehrlich, Donald Kennedy, Warren Washington, and Susan Joy Hassol. I thank leaders of the Youth Climate Movement, including Greta Thunberg, Alexandria Villaseñor, Luisa Neubauer, Ilana Cohen, Jerome Foster, and Jamie Margolin, for the inspiration they have provided.

I extend special thanks to colleagues who provided feedback and input, including Gavin Schmidt, Lee Kump, Matt Huber, Bill Ruddiman, Richard Alley, Dave Pollard, Rob DeConto, Matt Osman, Stefan Rahmstorf, and Matteo Willeit. And I offer thanks to my various colleagues and staff at the University of Pennsylvania who have made me feel so welcome there, including, among many others, President Liz Magill, Deans Steven Fluharty and John L. Jackson Jr., Katherine Unger Baillie, David Brainard, Joan A. Buccilli, Shannon

Christiansen, Bill Cohen, Cornelia Colijn, Nicholas Crivaro, Ezekiel Emanuel, Emily Falk, Joseph Francisco, David Goldsby, Vit Henisz, Kathleen Hall Jamieson, LaShawn R. Jefferson, Doug Jerolmack, Heather Kostick, Sarah E. Light, Irina Marinov, Bekezela Mbofana, Kathleen Morrison, Jennifer Pinto-Martin, Simon Richter, Vanessa Schipani, and Michael Weisberg.

I am greatly indebted to the various politicians on both ends of the political spectrum who stood up against powerful interests to support and defend me and other scientists against politically motivated attacks, and who have worked to advance the cause of an informed climate policy discourse. Among them are Sherwood Boehlert, Jerry Brown, Bob Bullard, Bob Casey Jr., Bill Clinton, Hillary Clinton, Peter Garrett, Al Gore, Mark Herring, Bob Inglis, Jay Inslee, Edward Markey, Terry McAuliffe, John McCain, Christine Milne, Jim Moran, Alexandria Ocasio-Cortez, Harry Reid, Bernie Sanders, Arnold Schwarzenegger, Arlen Specter, Malcolm Turnbull, Henry Waxman, Sheldon Whitehouse, and their various staff.

I also want to thank my agents Jodie Solomon, Rachel Vogel, and Suzi Jamil and the PublicAffairs crew, including my editor Colleen Lawrie, publicists Brooke Parsons and Miguel Cervantes, copyeditor Charlotte Byrnes, production editor Michelle Welsh-Horst, and jacket cover artist Pete Garceau, for all their hard work and support.

I wish to thank the various other friends, supporters, and colleagues past and present for their assistance, collaboration, friendship, and inspiration over the years, including John Abraham, Kylie Ahern, Ken Alex, Yoca Arditi-Rocha, Kurt Bardella, Ed Begley Jr., Andre Berger, Lew Blaustein, Doug Bostrom, Max Boykoff, Ray Bradley, Sir Richard Branson, Jonathan Brockopp, Bill Brune, James Byrne, Mike Cannon-Brookes, Elizabeth Carpino, Nick Carpino, Keya Chatterjee, Noam Chomsky, Kim Cobb, Ford Cochran, Michel Cochran, Julie Cole, John Collee, Leila Conners, John Cook, Jason Cronk, Jen Cronk, Michel Crucifix, Heidi Cullen, Hunter Cutting, Greg Dalton, Fred Damon, Kert Davies, Didier de Fontaine, Brendan Demelle, Andrew Dessler, Steve D'Hondt, Henry Diaz, Leonardo DiCaprio, Paulo D'Oderico, Pete Dominick, Andrea Dutton, Bill Easterling, Kerry Emanuel, Matt England, Howie Epstein, Jenni Evans, Morgan Fairchild, Thierry Fichefet, Chris Field, Frances

Fisher, Pete Fontaine, Josh Fox, Al Franken, Peter Frumhoff, Jose Fuentes, Andra Garner, Peter Garrett, Peter Gleick, Jeff Goodell, Amy Goodman, Hugues Goosse, Nellie Gorbea, David Graves, David Grinspoon, David Halpern, Thom Hartmann, David Haslingden, Susan Joy Hassol, Katharine Hayhoe, Tony Haymet, Megan Herbert, Bill Higgins, Michele Hollis, Rob Honeycutt, Ben Horton, Malcolm Hughes, Jan Jarrett, Paul Johansen, Phil Jones, Jim Kasting, Bill Keene, Sheril Kirshenbaum, Barbara Kiser, Johanna Köb, Jonathan Koomey, Miroslava Korenha, Kalee Kreider, Paul Krugman, Lauren Kurtz, Greg Laden, Chris Larson, Deb Lawrence, Tony Leiserowitz, Stephan Lewandowsky, Diccon Loxton, Ed Maibach, Scott Mandia, Joseph Marron, John Mashey, Francois Massonnet, Roger McConchie, Andrea McGimsey, Bill McKibben, Pete Meyers, Sonya Miller, Chris Mooney, John Morales, Granger Morgan, Ellen Mosely-Thompson, Leilani Munter, Ray Najjar, Giordano Nanni, Jeff Nesbit, Phil Newell, Gerald North, Dana Nuccitelli, Miriam O'Brien, Michael Oppenheimer, Naomi Oreskes, Tim Osborn, Jonathan Overpeck, Lisa Oxboel, Rajendra Pachauri, Blair Palese, David Paradice, Jeffrey Park, Rick Piltz, Phil Plait, James Powell, Stefan Rahmstorf, Cliff Rechtschaffen, Hank Reichman, Ann Reid, Catherine Reilly, James Renwick, Andy Revkin, Tom Richard, David Ritter, Alan Robock, Joe Romm, Lyndall Rowley, Mark Ruffalo, Scott Rutherford, Sasha Sagan, Barry Saltzman, Ben Santer, Julie Schmid, Gavin Schmidt, Steve Schneider, John Schwartz, Eugenie Scott, Joan Scott, Marshall Shepherd, Drew Shindell, Randy Showstack, Hank Shugart, David Silbert, Peter Sinclair, Michael Smerconish, Dave Smith, Jodi Solomon, Richard Somerville, Graham Spanier, Amanda Staudt, Eric Steig, Byron Steinman, David Stensrud, Nick Stokes, Sean Sublette, Larry Tanner, Jake Tapper, Lonnie Thompson, Sarah Thompson, Kim Tingley, Dave Titley, Lawrence Torcello, Kevin Trenberth, Fred Treyz, Katy Tur, Leah Tyrrell, Ana Unruh-Cohen, Jean-Pascal van Ypersele, Ali Velshi, Dave Verardo, Mikhail Verbitsky, David Vladeck, Nikki Vo, Bob Ward, Bud Ward, Bill Weir, Ray Weymann, Robert Wilcher, John B. Williams, Barbel Winkler, and Christopher Wright.

Figure Credits

Figure 1: Matthew B. Osman, Jessica E. Tierney, Jiang Zhu, Robert Tardif, Gregory J. Hakim, et al., "Globally Resolved Surface Temperatures Since the Last Glacial Maximum," *Nature* 599, 239–244 (2021), https://doi.org/10.1038/s41586-021-03984-4.

Figure 2: Courtesy of Ray Troll.

Figure 3: National Oceanographic and Atmospheric Administration (NOAA), "climateqa_hottest_ocean_temp_610.png," *Climate.gov*, February 9, 2021, https://www.climate.gov/media/11915. Graph by Hunter Allen and Michon Scott, using data from the NOAA National Climatic Data Center, courtesy of Carrie Morrill.

Figure 4: "File: Ocean circulation conveyor belt.jpg," *Wikipedia*, November 21, 2007, https://en.m.wikipedia.org/wiki/File:Ocean_circulation_conveyor_belt.jpg.

Figure 5: Christian Heinrich (Foto Feeling).

Figure 6: Lee R. Kump, James F. Kasting, and Robert G. Crane, *The Earth System* (New York: Pearson, 2004).

Figure 7: Courtesy of Bill Schopf.

Figure 8: Deborah Dardis and Pat Pendarvi, "Petroleum Is a Fossil Fuel," *Southeastern Louisiana University*, July 12, 2010, http://www2.southeastern.edu/orgs/oilspill/fossil.html. Modified from: Lucy E. Edwards and John Pojeta, Jr., "Fossils, Rocks, and Time," U.S. Geological Survey, U.S. Department of the Interior, 1993.

Figure 9: T. H. Torsvik, M. A. Smethurst, J. G. Meert, R. Van der Voo, W. S. McKerrow, et al., "Continental Break-up and Collision in the Neoproterozoic and Palaeozoic—A Tale of Baltica and Laurentia," *Earth-Science Reviews* 40(3–4), 229–258 (1996), https://doi.org/10.1016/0012-8252(96)00008-6. Copyright Elsevier.

Figure 10: USRA/Lunar and Planetary Institute.

Figure 11: Bruce Dorminey, "Did a Comet Fragment Kill the Dinosaurs? Not Likely, Say Researchers," *Forbes*, February 28, 2021, https://www.forbes.com/sites/brucedorminey/2021/02/28/did-a-comet-fragment-kill-the-dinosaurs-not-likely-say-researchers/. Illustration by David A. Kring.

Figure 12: BBC/John Sayer.

Figure 13: *Spherical Cow* by Piumadaquila.com, from Adobe Stock.

Figure 14: Photograph by the author.

Figure 15: *Mesohippus* by Heinrich Harder (1858–1935).

Figure 16: Adapted from M. Willeit, A. Ganopolski, R. Calov, and V. Brovkin, "Mid-Pleistocene Transition in Glacial Cycles Explained by Declining CO_2 and Regolith Removal," *Science Advances* 5(4), eaav7337 (2019), https://doi.org/10.1126/sciadv.aav7337. Courtesy of Matteo Willeit, Potsdam Institute for Climate Impact Research.

Figure 17: Lake Pit at La Brea Tar Pits, courtesy of the Natural History Museum of Los Angeles County.

Figure 18: Courtesy of Elijah Wolfson for *Time* magazine (Michael E. Mann, "'Widespread and Severe.' The Climate Crisis Is Here, but There's Still Time to Limit the Damage," August 9, 2021, https://time.com/6088531/ipcc-climate-report-hockey-stick-curve/).

Figure 19: Figure 2.28(b) from Sergey K. Gulev, Peter W. Thorne, Jinho Ahn, Frank J. Dentener, Catia M. Domingues, et al., "Changing State of the Climate System," in Valérie Masson-Delmotte, Panmao Zhai, Anna Pirani, Sarah L. Connors, Clotilde Péan, et al. (eds.), *Climate Change 2021: The Physical Science Basis*, Working Group I Contribution to the Sixth Assessment Report of the Intergovernmental Panel on Climate Change (Cambridge, UK: Cambridge University Press, 2021), 287–422, https://doi.org/10.1017/9781009157896.004.

Figure 20: Copyright © (2014) SCIENTIFIC AMERICAN, a Division of Nature American, Inc. All rights reserved.

Figure 21: From University of California–Santa Cruz, "High-Fidelity Record of Earth's Climate History Puts Current Changes in Context," *Phys.org*, September 10, 2020, https://phys.org/news/2020-09-high-fidelity-earth-climate-history-current.html. Courtesy of Thomas Westerhold.

Notes

Introduction

1. In this book, we will generally use the (almost exclusively now) American convention of measuring temperatures in degrees Fahrenheit (°F), though both units are given when dealing with quantities that have been scientifically defined in Celsius. Those in other countries can use the simple conversation that any change in temperature in degrees Fahrenheit must be multiplied by 0.56 to yield an equivalent change in degrees Celsius (°C).

2. Ira Flatow, "Truth, Deception, and the Myth of the One-Handed Scientist," *The Humanist*, October 18, 2012, https://thehumanist.com/magazine /november-december-2012/features/truth-deception-and-the-myth-of-the-one -handed-scientist/.

3. The term was coined by former *New York Times* scientist Andrew Revkin. See: Andrew C. Revkin, "Media Mania for a 'Front-Page Thought' on Climate," by *Dot Earth New York Times Blog*, December 14, 2007, https://archive .nytimes.com/dotearth.blogs.nytimes.com/2007/12/14/the-mania-for-a-front -page-thought-on-climate/.

Chapter 1: Our Moment Begins

1. Blythe A. Williams, "Effects of Climate Change on Primate Evolution in the Cenozoic," *Nature Education Knowledge* 7(1), 1 (2016).

2. M. E. Raymo and W. F. Ruddiman, "Tectonic Forcing of Late Cenozoic Climate," *Nature* 359, 117–122 (1992), https://doi.org/10.1038/359117a0.

3. See: James Zachos, Mark Pagani, Lisa Sloan, Ellen Thomas, and Katharina Billups, "Trends, Rhythms, and Aberrations in Global Climate 65 Ma to Present," *Science* 292, 686–693 (2001), https://doi.org/10.1126/science.1059412.

4. Tom Yulsman, "In the Blink of an Eye, We're Turning Back the Climatic Clock by 50 Million Years," *Discover Magazine*, December 14, 2018, https://www.discovermagazine.com/environment/in-the-blink-of-an-eye-were-turning-back-the-climatic-clock-by-50-million-years.

5. James C. Zachos, Gerald R. Dickens, and Richard E. Zeebe, "An Early Cenozoic Perspective on Greenhouse Warming and Carbon-Cycle Dynamics," *Nature* 451, 279–283 (2008), https://doi.org/10.1038/nature06588; R. Potts, "Environmental Hypotheses of Pliocene Human Evolution," in R. Bobe, Z. Alemseged, and A. K. Behrensmeyer (eds.), *Hominin Environments in the East African Pliocene: An Assessment of the Faunal Evidence* (Dordrecht: Springer, 2007), 25–49.

6. M. E. Raymo, "The Timing of Major Climate Terminations," *Paleoceanography and Paleoclimatology* 12, 577–585 (1997), https://doi.org/10.1029/97PA01169.

7. Nicholas R. Longrich, "When Did We Become Fully Human? What Fossils and DNA Tell Us About the Evolution of Modern Intelligence," *The Conversation*, September 9, 2020, https://theconversation.com/when-did-we-become-fully-human-what-fossils-and-dna-tell-us-about-the-evolution-of-modern-intelligence-143717.

8. Jason Daley, "Climate Change Likely Iced Neanderthals Out of Existence," *Smithsonian Magazine*, August 29, 2018, https://www.smithsonianmag.com/smart-news/modern-humans-didnt-kill-neanderthals-weather-did-180970167/.

9. John Noble Wilford, "When Humans Became Human," *New York Times*, February 26, 2002, https://www.nytimes.com/2002/02/26/science/when-humans-became-human.html.

10. James Trefil, "Evidence for a Flood," *Smithsonian Magazine*, April 1, 2000, https://www.smithsonianmag.com/science-nature/evidence-for-a-flood-102813115/; David R. Montgomery, "Biblical-Type Floods Are Real, and They're Absolutely Enormous," *Discover*, August 29, 2012, https://www.discovermagazine.com/planet-earth/biblical-type-floods-are-real-and-theyre-absolutely-enormous.

11. Julian B. Murton, Mark D. Bateman, Scott R. Dallimore, James T. Teller, and Zhirong Yang, "Identification of Younger Dryas Outburst Flood Path from Lake Agassiz to the Arctic Ocean," *Nature* 464, 740–743 (2010), https://doi.org/10.1038/nature08954.

12. Stefan Rahmstorf, Jason E. Box, Georg Feulner, Michael E. Mann, Alexander Robinson, et al., "Exceptional Twentieth-Century Slowdown in Atlantic Ocean Overturning Circulation," *Nature Climate Change* 5, 475–480 (2015), https://doi.org/10.1038/nclimate2554.

13. Conrad P. Kottak, *Window on Humanity: A Concise Introduction to Anthropology* (Boston: McGraw-Hill, 2005), 155–156.

14. Ofer Bar-Yosef, "Climatic Fluctuations and Early Farming in West and East Asia," *Current Anthropology* 52(S4), S175–S193 (2011), https://doi.org/10.1086/659784.

15. There is evidence of cultivation, for example, at the archeological site of Tell Qaramel, just north of modern-day Aleppo, dating back to 11,700 years ago; the site is home to the earliest known temples—two of them, in fact.

16. Bar-Yosef, "Climatic Fluctuations and Early Farming in West and East Asia."

17. Takuro Kobashi, Jeffrey P. Severinghaus, Edward J. Brook, Jean-Marc Barnola, and Alexi M. Grachev, "Precise Timing and Characterization of Abrupt Climate Change 8200 Years Ago from Air Trapped in Polar Ice," *Quaternary Science Reviews* 26(9–10), 1212–1222 (2007), https://doi.org/10.1016/j.quas cirev.2007.01.009; Harvey Weiss and Raymond S. Bradley, "What Drives Societal Collapse?" *Science* 291, 609–610 (2001), https://doi.org/10.1126 /science.1058775.

18. Daniel Hillel, *Rivers of Eden: The Struggle for Water and the Quest for Peace in the Middle East* (New York: Oxford University Press, 1994), 355; Weiss and Bradley, "What Drives Societal Collapse?"

19. Max Engel and Helmut Brückner, "Holocene Climate Variability of Mesopotamia and Its Impact on the History of Civilisation," in Eckart Ehlers and Katajun Amirpur (eds.), *Middle East and North Africa: Climate, Culture and Conflicts* (Leiden, The Netherlands: Brill, 2021), Chapter 3, https://doi .org/10.1163/9789004444973_005.

20. Heinz Wanner, Jürg Beer, Jonathan Bütikofer, Thomas J. Crowley, Ulrich Cubasch, et al., "Mid- to Late Holocene Climate Change: An Overview," *Quaternary Science Reviews* 27(19–20), 1791–1828 (2008), https://doi.org/10.1016 /j.quascirev.2008.06.013; Weiyi Sun, Bin Wang, Qiong Zhang, Deliang Chen, Guonian Lu, and Jian Liu, "Middle East Climate Response to the Saharan Vegetation Collapse During the Mid-Holocene," *Journal of Climate* 34(1), 229–242 (2021), https://doi.org/10.1175/JCLI-D-20-0317.1; Ian J. Orland, Feng He, Miryam Bar-Matthews, Guangshan Chen, Avner Ayalon, and John E. Kutzbach, "Resolving Seasonal Rainfall Changes in the Middle East During the Last Interglacial Period," *Proceedings of the National Academy of Sciences* 116(50), 24985–24990 (2019), https://doi.org/10.1073/pnas.1903139116.

21. R. E. Sojka, D. L. Bjorneberg, and J. A. Entry, "Irrigation: An Historical Perspective," in R. Lal (ed.), *Encyclopedia of Soil Science*, First Edition (New York: Marcel Dekker, Inc., 2002), 745–749.

22. H. Weiss, M.-A. Courty, W. Wetterstrom, F. Guichard, L. Senior, et al., "The Genesis and Collapse of Third Millennium North Mesopotamian Civilization," *Science* 261, 995–1004 (1993), https://doi.org/10.1126/science .261.5124.995; Weiss and Bradley, "What Drives Societal Collapse?"; Vasile Ersek, "How Climate Change Caused the World's First Ever Empire to Collapse," *The Conversation*, January 3, 2019, https://theconversation.com/how -climate-change-caused-the-worlds-first-ever-empire-to-collapse-109060.

23. Ersek, "How Climate Change Caused the World's First Ever Empire to Collapse"; Raymond S. Bradley and Jostein Bakke, "Is There Evidence for a

4.2 ka BP Event in the Northern North Atlantic Region?" *Climate of the Past* 15, 1665–1676 (2019), https://doi.org/10.5194/cp-15-1665-2019. Weiss sought out paleoclimate experts at Yale who might be able to provide confirmatory evidence of climate disruption when he was first pursuing this hypothesis during the early 1990s—I was one of them, which is how I first learned of this work.

24. The Indus Valley Civilization did dissipate several centuries later, in part because it had been dependent on trade with Mesopotamia, which was impacted by the fall of the Akkadian Empire: "What Happened to the Indus Civilisation?" *BBC Bitesize*, accessed February 17, 2022, https://www.bbc .co.uk/bitesize/topics/zxn3r82/articles/z8b987h. The Minoan civilization was destroyed by a volcanic eruption around 3500 BP. Several other Bronze Age civilizations collapsed shortly after 3200 BP. The notion that the Old Kingdom of Egypt underwent collapse has been contested by more recent research that suggests a relatively smooth transition from the old to new kingdoms: Andrew Lawler, "Did Egypt's Old Kingdom Die—or Simply Fade Away?" *National Geographic*, December 24, 2015, https://www.nationalgeographic.com/history /article/151224-egypt-climate-change-old-kingdom-archaeology.

25. For a number of alternative hypotheses, in addition to the monsoon /trade-wind hypothesis, see: Sarah M. White, A. Christina Ravelo, and Pratigya J. Polissar, "Dampened El Niño in the Early and Mid-Holocene Due to Insolation-Forced Warming/Deepening of the Thermocline," *Geophysical Research Letters* 45, 316–326 (2017), https://doi.org/10.1002/2017GL075433.

26. Northern Illinois University, "Archaeologists Shed New Light on Americas' Earliest Known Civilization," *Science Daily*, January 4, 2005, https://www .sciencedaily.com/releases/2005/01/050104112957.htm.

27. Lawler, "Did Egypt's Old Kingdom Die—or Simply Fade Away?"

28. See, for example: Kyle Harper, "6 Ways Climate Change and Disease Helped Topple the Roman Empire," *Vox*, November 4, 2017, https://www .vox.com/the-big-idea/2017/10/30/16568716/six-ways-climate-change-disease -toppled-roman-empire; Kyle Harper, "How Climate Change and Plague Helped Bring Down the Roman Empire," *Smithsonian Magazine*, December 19, 2017, https://www.smithsonianmag.com/science-nature/how-climate-change-and -disease-helped-fall-rome-180967591/.

29. See: Dagomar Degroot, Kevin Anchukaitis, Martin Bauch, Jakob Burnham, Fred Carnegy, et al., "Towards a Rigorous Understanding of Societal Responses to Climate Change," *Nature* 591, 539–550 (2021), https://doi .org/10.1038/s41586-021-03190-2; J. Luterbacher, J. P. Werner, J. E. Smerdon, L. Fernández-Donado, F. J. González-Rouco, et al., "European Summer Temperatures Since Roman Times," *Environmental Research Letters* 11, 024001 (2016), https://doi.org/10.1088/1748-9326/11/2/024001.

30. Drew T. Shindell, Gavin A. Schmidt, Michael E. Mann, David Rind, and Anne Waple, "Solar Forcing of Regional Climate Change During the Maunder Minimum," *Science* 294, 2149–2152 (2001), https://doi.org/10.1126 /science.1064363; Ilya G. Usoskin, "A History of Solar Activity over Millennia,"

Living Reviews in Solar Physics 10, 1 (2013), https://doi.org/10.12942/lrsp -2013-1.

31. Degroot et al., "Towards a Rigorous Understanding of Societal Responses to Climate Change."

32. Sarah Zielinski, "Plague Pandemic May Have Been Driven by Climate, Not Rats," *Smithsonian Magazine*, February 23, 2015, https://www.smith sonianmag.com/science-nature/plague-pandemic-may-have-been-driven-climate -not-rats-180954378/. The year 1453 CE, coincidentally, also saw one of the largest volcanic eruptions of the past 1000 years: the Kuwae eruption in Vanuatu in the tropical western Pacific Ocean. But there's no evidence that it played any role whatsoever in the fall of Constantinople that year. Coincidences do happen!

33. See: Michael E. Mann, Zhihua Zhang, Scott Rutherford, Raymond S. Bradley, Malcolm K. Hughes, et al., "Global Signatures and Dynamical Origins of the Little Ice Age and Medieval Climate Anomaly," *Science* 326, 1256–1260 (2009), https://doi.org/10.1126/science.1177303; Michael E. Mann, "Beyond the Hockey Stick: Climate Lessons from the Common Era," *Proceedings of the National Academy of Sciences* 118(39), e2112797118 (2021), https://doi .org/10.1073/pnas.2112797118.

34. Eli Kintisch, "Why Did Greenland's Vikings Disappear?" *Science*, November 10, 2016, https://www.science.org/content/article/why-did-greenland-s -vikings-disappear; George Dvorsky, "Over-Hunting Walruses Likely Forced Vikings to Abandon Greenland," *Gizmodo*, January 7, 2020, https://gizmodo .com/over-hunting-walruses-likely-forced-vikings-to-abandon-1840859102; Michael E. Mann, "Medieval Climatic Optimum," in Michael C. MacCracken and John S. Perry (eds.), *Encyclopedia of Global Environmental Change* (London: John Wiley and Sons Ltd., 2001), 514–516; Michael E. Mann, "Little Ice Age," in Michael C. MacCracken and John S. Perry (eds.), *Encyclopedia of Global Environmental Change* (London: John Wiley and Sons Ltd, 2001), 504–509.

35. Jeffrey S. Dean, William H. Doelle, and Janet D. Orcutt, "Adaptive Stress: Environment and Demography," in George J. Gumerman (ed.), *Themes in Southwest Prehistory* (Santa Fe: School for Advanced Research, 1993), 53–86; Robert L. Axtell, Joshua M. Epstein, Jeffrey S. Dean, George J. Gumerman, Alan C. Swedlund, et al., "Population Growth and Collapse in a Multiagent Model of the Kayenta Anasazi in Long House Valley," *Proceedings of the National Academy of Sciences* 99, 7275–7279 (2002), https://doi.org/10.1073 /pnas.092080799.

36. Degroot et al., "Towards a Rigorous Understanding of Societal Responses to Climate Change."

37. Mike Duncan, "Climate Chaos Helped Spark the French Revolution— and Holds a Dire Warning for Today," *Time*, October 20, 2021, https://time .com/6107671/french-revolution-history-climate/.

38. For a representative report, see: Michael Reilly, "'Little Ice Age' Hastened Fall of Aztecs, Incas," *NBC News*, December 22, 2008, https://www.nbc

news.com/id/wbna28353083; Charles C. Mann, *1491: New Revelations of the Americas Before Columbus* (New York: Vintage Books, 2006); Peter M. J. Douglas, Arthur A. Demarest, Mark Brenner, and Marcello A. Canuto, "Impacts of Climate Change on the Collapse of Lowland Maya Civilization," *Annual Review of Earth and Planetary Sciences* 44, 613–645 (2016), https://doi .org/10.1146/annurev-earth-060115-012512.

39. William F. Ruddiman, *Plows, Plagues, and Petroleum: How Humans Took Control of Climate* (Princeton: Princeton University Press, 2005).

40. Richard J. Blaustein, "William Ruddiman and the Ruddiman Hypothesis," *Humans and Nature*, September 27, 2015, https://humansandnature.org /william-ruddiman-and-the-ruddiman-hypothesis/.

41. Carl Sagan, *Billions and Billions: Thoughts on Life and Death at the Brink of the Millennium* (New York: Random House, 1997).

42. Hannah Miao, "Climate Change Is a Major Factor Behind Increased Migration at U.S. Southern Border, Experts Say," *CNBC*, April 18, 2021, https://www.cnbc.com/2021/04/18/us-mexico-border-climate-change-factor -behind-increased-migration.html.

43. Andrew Freedman, "The Worst Drought in 900 Years Helped Spark Syria's Civil War," *Mashable*, March 2, 2016, https://mashable.com/article /syria-drought-900-years-civil-war; Colin P. Kelley, Shahrzad Mohtadi, Mark A. Cane, Richard Seager, and Yochanan Kushnir, "Climate Change in the Fertile Crescent and Implications of the Recent Syrian Drought," *Proceedings of the National Academy of Sciences* 112, 3241–3246 (2015), https://doi.org/10.1073 /pnas.1421533112.

Chapter 2: Gaia and Medea

1. For an application of energy balance models, see: Michael E. Mann, "Earth Will Cross the Climate Danger Threshold by 2036," *Scientific American*, April 1, 2014, https://www.scientificamerican.com/article/earth-will-cross -the-climate-danger-threshold-by-2036/. An elementary discussion of the mathematics is described in this supplementary piece: Michael E. Mann, "Why Global Warming Will Cross a Dangerous Threshold in 2036," *Scientific American*, April 1, 2014, https://www.scientificamerican.com/article/mann-why-global -warming-will-cross-a-dangerous-threshold-in-2036/.

2. Carl Sagan and George Mullen, "Earth and Mars: Evolution of Atmospheres and Surface Temperatures," *Science* 177, 52–56 (1972), https://doi.org /10.1126/science.177.4043.52.

3. For a good background discussion, see: Lee R. Kump, James F. Kasting, and Robert G. Crane, *The Earth System*, Third Edition (San Francisco: Prentice Hall, 2010).

4. Lynn Margulis, "Gaia Is a Tough Bitch," in John Brockman (ed.), *The Third Culture: Beyond the Scientific Revolution* (New York: Touchstone, 1995), Chapter 7.

5. James E. Lovelock and Lynn Margulis, "Atmospheric Homeostasis by and for the Biosphere: The Gaia Hypothesis," *Tellus* 26(1–2), 2–10 (1974), https://doi.org/10.3402/tellusa.v26i1-2.9731.

6. John Postgate, "Gaia Gets Too Big for Her Boots," *New Scientist* 118(1607), 60 (1988), quoted in Michael Ruse, "Earth's Holy Fool?" *Aeon*, January 14, 2013, https://aeon.co/essays/gaia-why-some-scientists-think-its-a-nonsensical-fantasy.

7. Robert A. Berner, "Weathering, Plants, and the Long-Term Carbon Cycle," *Geochimica et Cosmochimica Acta* 56(8), 3225–3231 (1992), https://doi.org/10.1016/0016-7037(92)90300-8.

8. Readers can find a profile of Lovelock in the first edition of *Dire Predictions: Understanding Global Warming* (New York: Pearson, 2008) by my former Penn State Geosciences colleague Lee R. Kump and me. Kump was a collaborator of Lovelock's since the mid-1990s.

9. For a more technical description of Daisyworld, see: Kump et al., *The Earth System*. For a more comprehensive discussion provided by Lovelock, see: James Lovelock, *Healing Gaia: Practical Medicine for the Planet* (New York: Harmony Books, 1991).

10. D. A. Evans, N. J. Beukes, and J. L. Kirschvink, "Low-Latitude Glaciation in the Palaeoproterozoic Era," *Nature* 386, 262–266 (1997), https://doi.org/10.1038/386262a0; George E. Williams and Phillip W. Schmidt, "Paleomagnetism of the Paleoproterozoic Gowganda and Lorrain Formations, Ontario: Low Paleolatitude for Huronian Glaciation," *Earth and Planetary Science Letters* 153(3–4), 157–169 (1997), https://doi.org/10.1016/S0012-821X(97)00181-7. Early evidence for low-latitude tillites, for example, was provided by: W. B. Harland, "Critical Evidence for a Great Infra-Cambrian Glaciation," *Geologische Rundschau* 54(1), 45–61 (1964), https://doi.org/10.1007/BF01821169. Evidence for dropstones is discussed in: S. K. Donovan and R. K. Pickerill, "Dropstones: Their Origin and Significance: A Comment," *Palaeogeography, Palaeoclimatology, Palaeoecology* 131(1–2), 175–178 (1997), https://doi.org/10.1016/S0031-0182(96)00150-2. Evidence of glacial striations was provided by: P. A. Jensen and E. Wulff-Pedersen, "Glacial or Non-glacial Origin for the Bigganjargga Tillite, Finnmark, Northern Norway," *Geological Magazine* 133(2), 137–145 (1996), https://doi.org/10.1017/S0016756800008657; A. Bekker, A. J. Kaufman, J. A. Karhu, and K. A. Eriksson, "Evidence for Paleoproterozoic Cap Carbonates in North America," *Precambrian Research* 37, 167–206 (2005), https://doi.org/10.1016/j.precamres.2005.03.009.

11. Robert E. Kopp, Joseph L. Kirschvink, Isaac A. Hilburn, and Cody Z. Nash, "The Paleoproterozoic Snowball Earth: A Climate Disaster Triggered by the Evolution of Oxygenic Photosynthesis," *Proceedings of the National Academy of Sciences* 102(32), 11131–11136 (2005), https://doi.org/10.1073/pnas.0504878102. Some scientists have challenged this interpretation. Alcott and colleagues claim that this spike in oxygen could have occurred from natural long-term geochemical cycles: Lewis J. Alcott, Benjamin J. W. Mills, and Simon W. Poulton, "Stepwise Earth Oxygenation Is an Inherent Property of

Global Biogeochemical Cycling," *Science* 366, 1333–1337 (2019), https://doi
.org/10.1126/science.aax6459. Other more recent work continues to argue that
microbes indeed played the primary role: Haitao Shang, Daniel H. Rothman, and
Gregory P. Fournier, "Oxidative Metabolisms Catalyzed Earth's Oxygenation,"
Nature Communications 13, 1328 (2022), https://doi.org/10.1038/s41467-022
-28996-0. However, the prevailing thought is that oxygenic photosynthesis may
have evolved tens to hundreds of millions of years before the Great Oxidation
Event, but a large input of reducing gases from Earth's mantle likely delayed
the oxygen spike; see, for example: Lee R. Kump, "Hypothesized Link Between
Neoproterozoic Greening of the Land Surface and the Establishment of an
Oxygen-Rich Atmosphere," *Proceedings of the National Academy of Sciences*
111, 14062–14065 (2014), https://doi.org/10.1073/pnas.132149611. Recent
analyses of sulfur isotope data demonstrate that the glaciation followed the spike
in oxygen, ruling out the alternative hypothesis that the glaciation itself was the
reason for the Great Oxidation Event: Matthew R. Warke, Tommaso Di Rocco,
Aubrey L. Zerkle, Aivo Lepland, Anthony R. Prave, et al., "The Great Oxidation
Event Preceded a Paleoproterozoic 'Snowball Earth,'" *Proceedings of the National Academy of Sciences* 117, 13314–13320 (2020), https://doi.org/10.1073
/pnas.2003090117.

12. Peter Ward, *The Medea Hypothesis: Is Life on Earth Ultimately Self-
Destructive?* (Princeton: Princeton University Press, 2009).

13. Christopher McKay, "Thickness of Tropical Ice and Photosynthesis on
a Snowball Earth," *Geophysical Research Letters* 27(14), 2153–2156 (2000),
https://doi.org/10.1029/2000GL008525.

14. The cover of the January 2000 issue of *Scientific American* advertised
an article on the putative Neoproterozoic ice age as "Beyond ice ages: A star-
tling theory of our planet's frozen past": Paul F. Hoffman and Daniel P. Schrag,
"Snowball Earth," *Scientific American* 282(1), 68–75 (2000). See also: Paul
F. Hoffman, Alan J. Kaufman, Galen P. Halverson, and Daniel P. Schrag, "A
Neoproterozoic Snowball Earth," *Science* 281, 1342–1346 (1998), https://doi
.org/10.1126/science.281.5381.1342; Jakub Žárský, Vojtěch Žárský, Martin
Hanáček, and Viktor Žárský, "Cryogenian Glacial Habitats as a Plant Terres-
trialisation Cradle—The Origin of the Anydrophytes and Zygnematophyceae
Split," *Frontiers in Plant Science* 12, 735020 (2021), https://doi.org/10.3389
/fpls.2021.735020; Kump, "Hypothesized Link Between Neoproterozoic Greening of the Land Surface and the Establishment of an Oxygen-Rich Atmosphere."

15. See, for example: N. M. Chumakov, "A Problem of Total Glaciations on
the Earth in the Late Precambrian," *Stratigraphy and Geological Correlation* 16,
107–119 (2008), https://doi.org/10.1134/S0869593808020019; Philip A. Allen
and James L. Etienne, "Sedimentary Challenge to Snowball Earth," *Nature Geoscience* 1, 817–825 (2008), https://doi.org/10.1038/ngeo355; Nicholas Eyles
and Nicole Januszczak, "'Zipper-Rift': A Tectonic Model for Neoproterozoic
Glaciations During the Breakup of Rodinia after 750 Ma," *Earth-Science Reviews* 65(1–2), 1–73 (2004), https://doi.org/10.1016/S0012-8252(03)00080-1;

Jiasheng Wang, Ganqing Jiang, Shuhai Xiao, Qing Li, and Qing Wei, "Carbon Isotope Evidence for Widespread Methane Seeps in the ca. 635 Ma Doushantuo Cap Carbonate in South China," *Geology* 36(5), 347–350 (2008), https://doi.org/10.1130/G24513A.1; William T. Hyde, Thomas J. Crowley, Steven K. Baum, and W. Richard Peltier, "Neoproterozoic 'Snowball Earth' Simulations with a Coupled Climate/Ice-Sheet Model," *Nature* 405, 425–429 (2000), https://doi.org/10.1038/35013005; Daniel P. Schrag and Paul F. Hoffman, "Life, Geology and Snowball Earth," *Nature* 409, 306 (2001), https://doi.org/10.1038/35053170.

16. For a nice review, see: Kendal McGuffie and Ann Henderson-Sellers, *A Climate Modelling Primer* (New York: Wiley, 1987). See also: Georg Feulner, "The Faint Young Sun Problem," *Reviews of Geophysics* 50(2), RG2006 (2012), https://doi.org/10.1029/2011RG000375. For the original work by Budyko and Sellers, see: M. I. Budyko, "The Effect of Solar Radiation Variations on the Climate of the Earth," *Tellus* 21(5), 611–619 (1969); William D. Sellers, "A Global Climatic Model Based on the Energy Balance of the Earth-Atmosphere System," *Journal of Applied Meteorology and Climatology* 8(3), 392–400 (1969).

17. The term *Snowball Earth* was coined in: Joseph Kirschvink, "Late Proterozoic Low-Latitude Global Glaciation: The Snowball Earth," in J. William Schopf and Cornelis Klein (eds.), *The Proterozoic Biosphere: A Multidisciplinary Study* (New York: Cambridge University Press, 1992), 51–52.

18. Feulner, "The Faint Young Sun Problem."

19. M. I. Budyko, "The Future Climate," *Eos* 53(10), 868–874 (1972), https://doi.org/10.1029/EO053i010p00868.

20. Dirk Notz and Julienne Stroeve, "Observed Arctic Sea-Ice Loss Directly Follows Anthropogenic CO_2 Emission," *Science* 354, 747–750 (2016), https://doi.org/10.1126/science.aag2345.

Chapter 3: The Great Dying Wasn't So Great

1. See: Steven M. Holland, "Ordovician-Silurian Extinction," *Encyclopedia Britannica*, last updated June 1, 2020, https://www.britannica.com/science/Ordovician-Silurian-extinction. It is worth noting that the extinction event occurred in two stages, and other factors—including anoxia in the oceans—might have been implicated in the later stage.

2. Robert A. Berner, "GEOCARBSULF: A Combined Model for Phanerozoic Atmospheric O_2 and CO_2," *Geochimica et Cosmochimica Acta* 70, 5653–5664 (2006), https://doi.org/10.1016/j.gca.2005.11.032.

3. Jeffery T. Kiehl and Christine A. Shields, "Climate Simulation of the Latest Permian: Implications for Mass Extinction," *Geology* 33(9), 757–760 (2005), https://doi.org/10.1130/G21654.1; R. M. H. Smith, "Changing Fluvial Environments Across the Permian-Triassic Boundary in the Karoo Basin, South Africa and Possible Causes of Tetrapod Extinctions," *Palaeogeography, Palaeoclimatology, Palaeoecology* 117(1–2), 81–104 (1995), https://

doi.org/10.1016/0031-0182(94)00119-S; Robert A. Berner, "Examination of Hypotheses for the Permo–Triassic Boundary Extinction by Carbon Cycle Modeling," *Proceedings of the National Academy of Sciences* 99(7), 4172–4177 (2002), https://doi.org/10.1073/pnas.032095199.

4. Raymond B. Huey and Peter D. Ward, "Hypoxia, Global Warming, and Terrestrial Late Permian Extinctions," *Science* 308, 398–401 (2005), https://doi.org/10.1126/science.1108019; Robert J. Brocklehurst, Emma R. Schachner, Jonathan R. Codd, and William I. Sellers, "Respiratory Evolution in Archosaurs," *Philosophical Transactions of the Royal Society B* 375, 20190140 (1793), https://doi.org/10.1098/rstb.2019.0140.

5. See the MIT press releases describing the two studies: Jennifer Chu, "An Extinction in the Blink of an Eye," *MIT News*, February 10, 2014, https://news.mit.edu/2014/an-extinction-in-the-blink-of-an-eye-0210; Jennifer Chu, "Siberian Traps Likely Culprit for End-Permian Extinction," *MIT News*, September 16, 2015, https://news.mit.edu/2015/siberian-traps-end-permian-extinction-0916.

6. See: Ying Cui and Lee R. Kump, "Global Warming and the End-Permian Extinction Event: Proxy and Modeling Perspectives," *Earth-Science Reviews* 149, 5–22 (2015), https://doi.org/10.1016/J.Earscirev.2014.04.007.

7. See: Scripps Institution for Oceanography, "Mauna Loa and South Pole Isotopic ^{13}C Ratio," *Scripps CO$_2$ Program*, last updated January 2023, https://scrippsco2.ucsd.edu/graphics_gallery/isotopic_data/mauna_loa_and_south_pole_isotopic_c13_ratio.html.

8. Thure E. Cerling, "Carbon Dioxide in the Atmosphere: Evidence from Cenozoic and Mesozoic Paleosols," *American Journal of Science* 291, 377–400 (1991), https://doi.org/10.2475/ajs.291.4.377.

9. Yuyang Wu, Daoliang Chu, Jinnan Tong, Haijun Song, Jacopo Dal Corso, et al., "Six-Fold Increase of Atmospheric pCO$_2$ During the Permian–Triassic Mass Extinction," *Nature Communications* 12, 2137 (2021), https://doi.org/10.1038/s41467-021-22298-7.

10. See: Cui and Kump, "Global Warming and the End-Permian Extinction Event: Proxy and Modeling Perspectives." One complication here is that the current pattern of warming reflects the amplifying high-latitude effects of ice melt. It's possible that there was very little ice around during the late Permian. If that's true, then there would have been less high-latitude amplification of warming, and the adjustment factor would therefore be smaller, perhaps not much larger than one.

11. Wu et al., "Six-Fold Increase of Atmospheric pCO$_2$ During the Permian–Triassic Mass Extinction."

12. The original report is: Ad Hoc Study Group on Carbon Dioxide and Climate, *Carbon Dioxide and Climate: A Scientific Assessment* (Washington, DC: National Academy of Sciences, 1979). See also: Neville Nicholls, "40 Years Ago, Scientists Predicted Climate Change. And Hey, They Were Right," *The Conversation*, July 22, 2019, https://theconversation.com/40-years-ago-scientists-predicted-climate-change-and-hey-they-were-right-120502; Reto Knutti and

Gabriele C. Hegerl, "The Equilibrium Sensitivity of the Earth's Temperature to Radiation Changes," *Nature Geoscience* 1, 735–743 (2008), https://doi.org/10.1038/ngeo337.

13. Adam Morton, "Summer's Bushfires Released More Carbon Dioxide than Australia Does in a Year," *The Guardian*, April 21, 2020, https://www.theguardian.com/australia-news/2020/apr/21/summers-bushfires-released-more-carbon-dioxide-than-australia-does-in-a-year.

14. Daniel J. Lunt, Alan M. Haywood, Gavin A. Schmidt, Ulrich Salzmann, Paul J. Valdes, and Harry J. Dowsett, "Earth System Sensitivity Inferred from Pliocene Modelling and Data," *Nature Geoscience* 3, 60–64 (2010), https://doi.org/10.1038/ngeo706.

15. Dana L. Royer, Robert A. Berner, and Jeffrey Park, "Climate Sensitivity Constrained by CO_2 Concentrations over the Past 420 Million Years," *Nature* 446, 530–532 (2007), https://doi.org/10.1038/nature05699.

16. See: Michael E. Mann, "Beyond the Hockey Stick: Climate Lessons from the Common Era," *Proceedings of the National Academy of Sciences* 118(39), e2112797118, https://doi.org/10.1073/pnas.2112797118, 2021.

17. See: M. O. Clarkson, S. A. Kasemann, R. A. Wood, T. M. Lenton, S. J. Daines, et al., "Ocean Acidification and the Permo-Triassic Mass Extinction," *Science* 348, 229–232 (2015), https://doi.org/10.1126/science.aaa0193.

18. Wu et al., "Six-Fold Increase of Atmospheric pCO_2 During the Permian–Triassic Mass Extinction."

19. Andrew H. Knoll, Richard K. Bambach, Jonathan L. Payne, Sara Pruss, and Woodward W. Fischer, "Paleophysiology and End-Permian Mass Extinction," *Earth and Planetary Science Letters* 256(3–4), 295–313 (2007), https://doi.org/10.1016/j.epsl.2007.02.018.

20. Guancheng Li, Lijing Cheng, Jiang Zhu, Kevin E. Trenberth, Michael E. Mann, and John P. Abraham, "Increasing Ocean Stratification over the Past Half-Century," *Nature Climate Change* 10, 1116–1123 (2020), https://doi.org/10.1038/s41558-020-00918-2. For a summary of the findings from the article, see: Michael E. Mann, "The Oceans Appear to Be Stabilizing. Here's Why It's Very Bad News," *Newsweek*, September 28, 2020, https://www.newsweek.com/climate-change-oceans-stabilizing-1534512.

21. Michael R. Rampino and Ken Caldeira, "Major Perturbation of Ocean Chemistry and a 'Strangelove Ocean' After the End-Permian Mass Extinction," *Terra Nova* 17(6), 554–559 (2005), https://doi.org/10.1111/j.1365-3121.2005.00648.x.

22. Lee R. Kump, Alexander Pavlov, and Michael A. Arthur, "Massive Release of Hydrogen Sulfide to the Surface Ocean and Atmosphere During Intervals of Oceanic Anoxia," *Geology* 33(5), 397–400 (2005), https://doi.org/10.1130/G21295.1.

23. Arthur Capet, Emil V. Stanev, Jean-Marie Beckers, James W. Murray, and Marilaure Grégoire, "Decline of the Black Sea Oxygen Inventory," *Biogeosciences* 13(4), 1287–1297 (2016), https://doi.org/10.5194/bg-13-1287-2016;

Peter Schwartzstein, "The Black Sea Is Dying, and War Might Push it Over the Edge," *Smithsonian Magazine*, May 11, 2016, https://www.smithsonianmag.com/science-nature/black-sea-dying-and-war-might-push-it-over-edge-180959053/.

24. Henk Visscher, Cindy V. Looy, Margaret E. Collinson, Henk Brinkhuis, Johanna H. A. van Konijnenburg-van Cittert, et al., "Environmental Mutagenesis During the End-Permian Ecological Crisis," *Proceedings of the National Academy of Sciences* 101, 12952–12956 (2004), https://doi.org/10.1073/pnas.0404472101.

25. "Global Warming Led to Climatic Hydrogen Sulfide and Permian Extinction," Penn State press release, February 21, 2005, https://www.psu.edu/news/research/story/global-warming-led-climatic-hydrogen-sulfide-and-permian-extinction/.

26. John Quiggin, "Australia Has Stalled for Too Long, We Need to Quit Fossil Fuels Now," *Gizmodo*, April 7, 2022, https://www.gizmodo.com.au/2022/04/australia-has-stalled-for-too-long-we-need-to-quit-fossil-fuels-now/.

27. Andrew Freedman, "Australia Fires: Yearly Greenhouse Gas Emissions Nearly Double Due to Historic Blazes," *The Independent*, January 25, 2020, https://www.independent.co.uk/news/world/australasia/australia-fires-greenhouse-gas-emissions-climate-crisis-fossil-fuel-a9301396.html; Laura Millan Lombrana, Hayley Warren, and Akshat Rathi, "Measuring the Carbon-Dioxide Cost of Last Year's Worldwide Wildfires," *Bloomberg*, February 10, 2020, https://www.bloomberg.com/graphics/2020-fire-emissions; Rhett A. Butler, "Amazon Destruction," *Mongabay*, November 23, 2021, https://rainforests.mongabay.com/amazon/amazon_destruction.html; Fiona Harvey, "Tropical Forests Losing Their Ability to Absorb Carbon, Study Finds," *The Guardian*, March 4, 2020, https://www.theguardian.com/environment/2020/mar/04/tropical-forests-losing-their-ability-to-absorb-carbon-study-finds; Wannes Hubau, Simon L. Lewis, Oliver L. Phillips, Kofi Affum-Baffoe, Hans Beeckman, et al., "Asynchronous Carbon Sink Saturation in African and Amazonian Tropical Forests," *Nature* 579, 80–87 (2020), https://doi.org/10.1038/s41586-020-2035-0.

28. See: Scripps Institution for Oceanography, "Scripps O_2 Global Oxygen Measurements," *Scripps O_2 Program*, accessed April 9, 2022, https://scrippso2.ucsd.edu/.

29. "Global Warming Led to Climatic Hydrogen Sulfide and Permian Extinction," Penn State press release; Andreas Oschlies, Peter Brandt, Lothar Stramma, and Sunke Schmidtko, "Drivers and Mechanisms of Ocean Deoxygenation," *Nature Geoscience* 11, 467–473 (2018), https://doi.org/10.1038/s41561-018-0152-2.

30. A nice review is provided by the National Oceanic and Atmospheric Administration: "Understanding Ocean Acidification," *NOAA Fisheries*, accessed April 9, 2022, https://www.fisheries.noaa.gov/insight/understanding-ocean-acidification. See also, for example, this report from the International Union for Conservation of Nature: "Latin American and Caribbean Countries Threatened by Rising Ocean Acidity, Experts Warn," *IUCN*, April 3, 2018, https://

www.iucn.org/news/secretariat/201804/latin-american-and-caribbean-countries
-threatened-rising-ocean-acidity-experts-warn.

31. O. Hoegh-Guldberg, P. J. Mumby, A. J. Hooten, R. S. Steneck, P. Green-field, et al., "Coral Reefs Under Rapid Climate Change and Ocean Acidification," *Science* 318, 1737–1742 (2007), https://doi.org/10.1126/science.1152509.

32. Kelsey Piper, "When the World Actually Solved an Environmental Crisis," *Vox*, October 3, 2021, https://web.archive.org/web/20211003120318, /https://www.vox.com/future-perfect/22686105/future-of-life-ozone-hole -environmental-crisis.

33. See, for example: Ian Johnston, "Earth's Worst-Ever Mass Extinction of Life Holds 'Apocalyptic' Warning About Climate Change, Say Scientists," *The Inde-pendent*, March 24, 2017, https://www.independent.co.uk/climate-change/news /earth-permian-mass-extinction-apocalypse-warning-climate-change-frozen -methane-a7648006.html; Dana Nuccitelli, "There are Genuine Climate Alarm-ists, but They're Not in the Same League as Deniers," *The Guardian*, July 9, 2018, https://www.theguardian.com/environment/climate-consensus-97-per-cent /2018/jul/09/there-are-genuine-climate-alarmists-but-theyre-not-in-the-same -league-as-deniers; Timothy Gardner, "Global Methane Emissions Rising Due to Oil and Gas, Agriculture—Studies," *Reuters*, July 14, 2020, https://www .reuters.com/article/us-climate-change-methane/global-methane-emissions -rising-due-to-oil-and-gas-agriculture-studies-idUSKCN24F2X8.

Chapter 4: Mighty Brontosaurus

1. Luis W. Alvarez, Walter Alvarez, Frank Asaro, and Helen V. Michel, "Ex-traterrestrial Cause for the Cretaceous-Tertiary Extinction," *Science* 208, 1095–1108 (1980), https://doi.org/10.1126/science.208.4448.1095.

2. The one exception is the semiterrestrial crocodile. For a comprehensive review, see: Peter Schulte, Laia Alegret, Ignacio Arenillas, José A. Arz, Penny J. Barton, et al., "The Chicxulub Asteroid Impact and Mass Extinction at the Cretaceous-Paleogene Boundary," *Science* 327(5970), 1214–1218 (2010), https://doi.org/10.1126/science.1177265.

3. "Bolide," *Wikipedia*, accessed April 25, 2022, https://en.wikipedia.org /wiki/Bolide; Alan R. Hildebrand and William V. Boynton, "Proximal Cretaceous-Tertiary Boundary Impact Deposits in the Caribbean," *Science* 248(4957), 843–847 (1990), https://doi.org/10.1126/science.248.4957.843.

4. Nicholas St. Fleur, "Drilling into the Chicxulub Crater, Ground Zero of the Dinosaur Extinction," *New York Times*, November 17, 2016, https://www .nytimes.com/2016/11/18/science/chicxulub-crater-dinosaur-extinction.html; Daphne Leprince-Ringuet, "Supercomputer Simulates the Impact of the Aster-oid that Wiped out Dinosaurs," *ZDNET*, May 26, 2020, https://www.zdnet.com /article/supercomputer-simulates-the-impact-of-the-asteroid-that-wiped-out -dinosaurs/; David Kindy, "Mile-High Tsunami Caused by Dinosaur-Killing Asteroid Left Behind Towering 'Megaripples,'" *Smithsonian Magazine*, July 19,

2021, https://www.smithsonianmag.com/smart-news/mile-high-tsunami-caused-dinosaur-killing-asteroid-left-behind-towering-megaripples-180978229/.

5. Jonathan Amos, "Tanis: Fossil Found of Dinosaur Killed in Asteroid Strike, Scientists Claim," *BBC*, April 6, 2022, https://www.bbc.com/news/science-environment-61013740.

6. Nola Taylor Redd, "After the Dinosaur-Killing Impact, Soot Played a Remarkable Role in Extinction," *Smithsonian Magazine*, April 27, 2020, https://www.smithsonianmag.com/science-nature/soot-dinosaur-impact-180974708/.

7. See, for example: Elizabeth Kolbert, *The Sixth Extinction: An Unnatural History* (London: Bloomsbury, 2014).

8. The film, as I recount in *The Hockey Stick and the Climate Wars* (New York: Columbia University Press, 2012), so impacted me that I ended up coding a self-learning tic-tac-toe program in school next fall.

9. Ronald Reagan, "White House Diaries: Monday, October 10, 1983," *The Ronald Reagan Presidential Foundation and Institute*, accessed September 10, 2019, https://www.reaganfoundation.org/ronald-reagan/white-house-diaries/diary-entry-10101983/.

10. See: "Carl," *BabyCentre*, accessed May 1, 2022, https://www.babycentre.co.uk/babyname/1005569/carl; Polly Logan-Banks, "Number 7 Numerology," *BabyCentre*, accessed May 1, 2022, https://www.babycentre.co.uk/a25017039/number-7-in-numerology. I feel compelled to point out that, as apt as this description might seem of Carl Sagan, he would be among the first to dismiss such superstitious assessments. And as a person of science, I would have to side with him. But I also feel compelled to note that Carl Jung might have embraced this ostensible additional example of synchronicity.

11. R. P. Turco, O. B. Toon, T. P. Ackerman, J. B. Pollack, and C. Sagan, "Nuclear Winter: Global Atmospheric Consequences of Nuclear War," *Science* 222, 1283–1292 (1983), https://doi.org/10.1126/science.222.4630.1283; for a comprehensive review of the so-called TTAPS article, see: Naomi Oreskes and Erik M. Conway, *Merchants of Doubt: How a Handful of Scientists Obscured the Truth on Issues from Tobacco Smoke to Global Warming* (New York: Bloomsbury, 2010).

12. "President Obama Names Harvard Physicist John P. Holdren as Science Adviser," *APS News*, February 2009, https://www.aps.org/publications/apsnews/200902/holdren.cfm; Paul R. Ehrlich, Anne H. Ehrlich, and John P. Holdren, *Ecoscience: Population, Resources, Environment* (San Francisco: Freeman, 1977); Paul R. Ehrlich, John Harte, Mark A. Harwell, Peter H. Raven, Carl Sagan, et al., "Long-Term Biological Consequences of Nuclear War," *Science* 222, 1293–1300 (1983), https://doi.org/10.1126/science.6658451; Paul J. Crutzen and John W. Birks, "The Atmosphere After a Nuclear War: Twilight at Noon," *Ambio* 11(2/3), 114–125 (1982).

13. "Walking in Your Footsteps," *The Police Wiki*, accessed September 17, 2021, http://www.thepolicewiki.org/Police_wiki/index.php?title=Walking_In_Your

_Footsteps. It is fair to quibble a bit with the lyrics, incidentally. We might take note, for example, that *Brontosaurus* was not a casualty of the K-Pg impact event. It actually went extinct seventy-nine million years earlier, in the late Jurassic period. Lest that seem like a minor dating error, let us note that more time elapsed between the demise of *Brontosaurus* and the occurrence of the K-Pg extinction event than has elapsed since the K-Pg event. While we're quibbling about dating, the K-Pg event was sixty-six million years ago, not the "fifty million years" mentioned in the lyrics. *Brontosaurus*, finally, wasn't actually even a "thing" by 1982. In the 1970s, it was discovered to, in reality, be the same species as the earlier-named species *Apatosaurus*. When naming dinosaurs, it's first come, first served. Paleontologists, following the rule book, had to cancel poor *Brontosaurus*. But an engaging song and an apt metaphor are deserving of some degree of leniency and poetic license, and we should forgive these minor transgressions.

14. Ed Regis, "The Doomslayer," *Wired*, February 1, 1997, https://www.wired.com/1997/02/the-doomslayer-2/; "World Scientists' Warning to Humanity," *Union of Concerned Scientists*, July 16, 1992, https://www.ucsusa.org/resources/1992-world-scientists-warning-humanity; "A Joint Statement by Fifty-Eight of the World's Scientific Academies," *Population Summit of the World's Scientific Academies*, 1993, http://www.nap.edu/openbook.php?record_id=9148&page=R2.

15. I've discussed the affair at length in: Michael E. Mann, *The Hockey Stick and the Climate Wars: Dispatches from the Front Lines* (New York: Columbia University Press, 2012).

16. I had the pleasure of participating in a one-on-one author conversation event for my last book, *The New Climate War*, with Sasha Sagan for the Brooklyn Public Library on February 17, 2021. It is archived here: BPLvideos, "Green Series: Michael Mann on 'The New Climate War' with Sasha Sagan" (video), *YouTube*, February 24, 2021, https://www.youtube.com/watch?v=nWbFnORuI4s.

17. Johnny Carson (host), *The Tonight Show*, episode 3885, aired March 2, 1978 on NBC; Matthew R. Francis, "When Carl Sagan Warned the World About Nuclear Winter," *Smithsonian Magazine*, November 15, 2017, https://www.smithsonianmag.com/science-nature/when-carl-sagan-warned-world-about-nuclear-winter-180967198/.

18. See, for example: Philip Ball, "Lessons from Cold Fusion, 30 Years On," *Nature*, May 27, 2019, https://www.nature.com/articles/d41586-019-01673-x.

19. Associated Press, "Hansen and Hannah Arrested in West Virginia Mining Protest," *The Guardian*, June 24, 2009, https://www.theguardian.com/environment/2009/jun/24/james-hansen-daryl-hannah-mining-protest; Jeanna Bryner, "NASA Climate Scientist Arrested in Pipeline Protest," *Live Science*, February 13, 2013, https://www.livescience.com/27117-nasa-climate-scientist-arrest

.html; Ethan Freedman, "'It's Critical the Message Makes It to the Mainstream': NASA Climate Scientist Speaks on His Tearful Protest," *The Independent*, April 17, 2022, https://www.independent.co.uk/climate-change/news/protest -nasa-scientist-rebellion-b2059788.html.

20. See: Keay Davidson, *Carl Sagan: A Life* (New York: John Wiley & Sons, 2000).

21. J. Hansen, D. Johnson, A. Lacis, S. Lebedeff, P. Lee, et al., "Climate Impact of Increasing Atmospheric Carbon Dioxide," *Science* 213, 957–966 (1981), https:// doi.org/10.1126/science.213.4511.957; Benjamin Franta, "Shell and Exxon's Secret 1980s Climate Change Warnings," *The Guardian*, September 19, 2018, https://www.theguardian.com/environment/climate-consensus-97-per-cent/ 2018/sep/19/shell-and-exxons-secret-1980s-climate-change-warnings.

22. See: U.S. National Academy of Sciences, *Understanding* Climate Change: A *Program for Action* (Washington, DC: National Academies Press, 1975); Thomas C. Peterson, William M. Connolley, and John Fleck, "The Myth of the 1970s Global Cooling Scientific Consensus," *Bulletin of the American Meteorological Society* 89(9), 1325–1338 (2008), https://doi.org/10.1175/2008BAMS2370.1; S. I. Rasool and S. H. Schneider, "Atmospheric Carbon Dioxide and Aerosols: Effects of Large Increases on Global Climate," *Science* 173, 138–141 (1971), https://doi.org/10.1126/science.173.3992.138.

23. A good example is this extremely misleading commentary: Conrad Black, "The Moving Targets of the Climate Change Movement," *The Hill*, March 25, 2021, https://thehill.com/opinion/energy-environment/544472-the-moving -targets-of-the-climate-change-movement/.

24. Starley L. Thompson and Stephen H. Schneider, "Nuclear Winter Re-appraised," *Foreign Affairs*, Summer 1986, https://www.foreignaffairs.com /articles/1986-06-01/nuclear-winter-reappraised; Owen Jarus, "The 9 Most Powerful Nuclear Weapon Explosions," *Live Science*, March 23, 2022, https:// www.livescience.com/most-powerful-nuclear-explosions; Curt Covey, Stephen H. Schneider, and Starley L. Thompson, "Global Atmospheric Effects of Massive Smoke Injections from a Nuclear War: Results from General Circulation Model Simulations," *Nature* 308, 21–25 (1984), https://doi.org/10.1038/308021a0. See also: S. L. Thompson, V. V. Aleksandrov, G. L. Stenchikov, S. H. Schneider, C. Covey, and R. M. Chervin, "Global Climatic Consequences of Nuclear War: Simulations with Three Dimensional Models," *Ambio* 13(4), 236–243 (1984).

25. Malcolm W. Browne, "Nuclear Winter Theorists Pull Back," *New York Times*, January 23, 1990, https://www.nytimes.com/1990/01/23/science/nuclear -winter-theorists-pull-back.html; R. P. Turco, O. B. Toon, T. P. Ackerman, J. B. Pollack, and C. Sagan, "Climate and Smoke: An Appraisal of Nuclear Winter," *Science* 247, 166–176 (1990), https://doi.org/10.1126/science.11538069.

26. Alan Robock, "Nuclear Winter Is a Real and Present Danger," *Nature* 473, 275–275 (2011), https://doi.org/10.1038/473275a.

27. See: Mann, *The Hockey Stick and the Climate Wars*.

28. William F. Buckley Jr., "The Specter of Nuclear War," *The Washington Post*, April 22, 1985; Matthew R. Francis, "When Carl Sagan Warned the World About Nuclear Winter," *Smithsonian Magazine*, November 15, 2017, https://www.smithsonianmag.com/science-nature/when-carl-sagan-warned -world-about-nuclear-winter-180967198/.

29. See: Oreskes and Conway, *Merchants of Doubt*; Ross Gelbspan, *The Heat Is On: The Climate Crisis, The Cover-up, The Prescription* (New York: Basic Books, 1997); James Hoggan, with Richard Littlemore, *Climate Cover-Up: The Crusade to Deny Global Warming* (Vancouver: Greystone, 2009); Mann, *The Hockey Stick and the Climate Wars*.

30. S. Fred Singer, "Is the 'Nuclear Winter' Real?" *Nature* 310, 625 (1984), https://doi.org/10.1038/310625a0; Jill Lepore, "The Atomic Origins of Climate Science: How Arguments About Nuclear Weapons Shaped the Debate over Global Warming," *New Yorker*, January 22, 2017, https://www.newyorker.com /magazine/2017/01/30/the-atomic-origins-of-climate-science; Singer, "Is the 'Nuclear Winter' Real?"; S. Fred Singer, "On a 'Nuclear Winter'" (letter), *Science* 227, 356 (1985), https://doi.org/10.1126/science.227.4685.356.a.

31. For a detailed account of these individuals, including their backgrounds and the history of how they became involved in industry-funded public relations campaigns including climate change denial, see: Oreskes and Conway, *Merchants of Doubt*; Davidson, *Carl Sagan*; Mann, *The Hockey Stick and the Climate Wars*.

32. Oreskes and Conway, *Merchants of Doubt*, Chapter 2.

33. Lepore, "The Atomic Origins of Climate Science"; Edward Teller, "Widespread After-Effects of Nuclear War," *Nature* 310, 621–624 (1984), https://doi .org/10.1038/310621a0.

34. See: Charles Schwartz, "Defying the Laws of Physics: A Professor Takes a Stand Against Star Wars," *Sojourners*, May 1987, https://sojo.net/magazine /may-1987/defying-laws-physics-professor-takes-stand-against-star-wars.

35. Sharon Begley, "The Truth About Denial," *Newsweek*, August 13, 2007.

36. Robock, "Nuclear Winter Is a Real and Present Danger."

37. The moderated debate between Patrick Michaels and Alan Robock took place on July 15, 1997, during The Costs of Kyoto conference held at the Competitive Enterprise Institute in Washington, DC. A transcript is available at *Climate Files*: https://www.climatefiles.com/deniers/patrick-michaels-collection /1997-cei-debate-climate-change-patrick-michaels-alan-robock/.

38. Carl Sagan, "With Science on Our Side," *The Washington Post*, January 9, 1994, https://www.washingtonpost.com/archive/entertainment/books /1994/01/09/with-science-on-our-side/9e5d2141-9d53-4b4b-aa0f-7a6a0fa ff845/; Alfio Alessandro Chiarenza, Alexander Farnsworth, Philip D. Mannion, Daniel J. Lunt, Paul J. Valdes, et al., "Asteroid Impact, Not Volcanism, Caused the End-Cretaceous Dinosaur Extinction," *Proceedings of the National*

Academy of Sciences 117(29), 17084–17093 (2020), https://doi.org/10.1073/pnas .2006087117; Paul R. Renne, Courtney J. Sprain, Mark A. Richards, Stephen Self, Loÿc Vanderkluysen, and Kanchan Pande, "State Shift in Deccan Volcanism at the Cretaceous-Paleogene Boundary, Possibly Induced by Impact," *Science* 350(6256), 76–78 (2015), https://doi.org/10.1126/science.aac7549.

39. Manabu Sakamoto, Michael J. Benton, and Chris Venditti, "Dinosaurs in Decline Tens of Millions of Years Before Their Final Extinction," *Proceedings of the National Academy of Sciences* 113(18), 5036–5040 (2016), https://doi .org/10.1073/pnas.1521478113.

40. Wallace S. Broecker and Tsung-Hung Peng, *Tracers in the Sea* (Palisades, NY: Lamont-Doherty Geological Observatory, 1982).

41. Broecker and Peng, *Tracers in the Sea.*

42. Steven D'Hondt, Percy Donaghay, James C. Zachos, Danielle Luttenberg, and Matthias Lindinger, "Organic Carbon Fluxes and Ecological Recovery from the Cretaceous-Tertiary Mass Extinction," *Science* 282, 276–279 (1998), https:// doi.org/10.1126/science.282.5387.276.

43. J. Brad Adams, Michael E. Mann, and Steven D'Hondt, "The Cretaceous-Tertiary Extinction: Modeling Carbon Flux and Ecological Response," *Paleoceanography and Paleoclimatology* 19, PA1002 (2004), https://doi.org/10.1029 /2002PA000849.

44. Jim Shelton, "Mystery Solved: Ocean Acidity in the Last Mass Extinction," *Yale University News*, October 21, 2019, https://news.yale.edu/2019/10/21 /mystery-solved-ocean-acidity-last-mass-extinction; Michael J. Henehan, Andy Ridgwell, Ellen Thomas, Shuang Zhang, Laia Alegret, et al., "Rapid Ocean Acidification and Protracted Earth System Recovery Followed the End-Cretaceous Chicxulub Impact," *Proceedings of the National Academy of Sciences* 116(45), 22500–22504 (2019), https://doi.org/10.1073/pnas.1905989116.

45. Henehan et al., "Rapid Ocean Acidification and Protracted Earth System Recovery."

46. T. J. Raphael and Jack D'Isidoro, "How the Threat of Nuclear Winter Changed the Cold War," *The World*, April 5, 2016, https://theworld.org /stories/2016-04-05/how-threat-nuclear-winter-changed-cold-war.

47. Henry Jacoby, Benjamin Santer, Gary Yohe, and Richard Richels, "Fighting Climate Change in a Fragmented World," *The Hill*, May 7, 2022, https://thehill .com/opinion/energy-environment/3479916-fighting-climate-change-in-a-frag mented-world/; Michael E. Mann, *The New Climate War: The Fight to Take Back Our Planet* (New York: PublicAffairs, 2020).

48. Gordon N. Inglis, Fran Bragg, Natalie J. Burls, Margot J. Cramwinckel, David Evans, et al., "Global Mean Surface Temperature and Climate Sensitivity of the Early Eocene Climatic Optimum (EECO), Paleocene–Eocene Thermal Maximum (PETM), and Latest Paleocene," *Climate of the Past* 16(5), 1953–1968 (2020), https://doi.org/10.5194/cp-16-1953-2020.

Chapter 5: Hothouse Earth

1. Tom Dunkley Jones, Daniel J. Lunt, Daniela N. Schmidt, Andy Ridgwell, Appy Sluijs, et al., "Climate Model and Proxy Data Constraints on Ocean Warming Across the Paleocene–Eocene Thermal Maximum," *Earth-Science Reviews* 125, 123–145 (2013), https://doi.org/10.1016/j.earscirev.2013.07.004.

2. See: Francesca A. McInerney and Scott L. Wing, "The Paleocene-Eocene Thermal Maximum: A Perturbation of Carbon Cycle, Climate, and Biosphere with Implications for the Future," *Annual Review of Earth and Planetary Sciences* 39, 489–516 (2011), https://doi.org/10.1146/annurev-earth-040610-133431; Aradhna K. Tripati and Henry Elderfield, "Abrupt Hydrographic Changes in the Equatorial Pacific and Subtropical Atlantic from Foraminiferal Mg/Ca Indicate Greenhouse Origin for the Thermal Maximum at the Paleocene-Eocene Boundary," *Geochemistry, Geophysics, Geosystems* 5(2), Q02006 (2004), https://doi.org/10.1029/2003GC000631.

3. David A. Carozza, Lawrence A. Mysak, and Gavin A. Schmidt, "Methane and Environmental Change During the Paleocene-Eocene Thermal Maximum (PETM): Modeling the PETM Onset as a Two-Stage Event," *Geophysical Research Letters* 38(5), L05702 (2011), https://doi.org/10.1029/2010GL046038.

4. Alexander Gehler, Philip D. Gingerich, and Andreas Pack, "Temperature and Atmospheric CO_2 Concentration Estimates Through the PETM Using Triple Oxygen Isotope Analysis of Mammalian Bioapatite," *Proceedings of the National Academy of Sciences* 113(28), 7739–7744 (2016), https://doi.org/10.1073/pnas.1518116113.

5. Richard E. Zeebe, Andy Ridgwell, and James C. Zachos, "Anthropogenic Carbon Release Rate Unprecedented During the Past 66 Million Years," *Nature Geoscience* 9, 325–329 (2016), https://doi.org/10.1038/ngeo2681. See also: Andrew Freedman, "Carbon Dioxide Is Rising at Its Fastest Rate in 66 Million Years," *Mashable*, March 21, 2016, https://mashable.com/article/co2-fastest-66-million-years.

6. Christophe McGlade and Paul Ekins, "The Geographical Distribution of Fossil Fuels Unused When Limiting Global Warming to 2°C," *Nature* 517, 187–190 (2015), https://doi.org/10.1038/nature14016.

7. Thomas A. Stidham and Jaelyn J. Eberle, "The Palaeobiology of High Latitude Birds from the Early Eocene Greenhouse of Ellesmere Island, Arctic Canada," *Scientific Reports* 6, 20912 (2016), https://doi.org/10.1038/srep20912; Joost Frieling, Holger Gebhardt, Matthew Huber, Olabisi A. Adekeye, Samuel O. Akande, et al., "Extreme Warmth and Heat-Stressed Plankton in the Tropics During the Paleocene-Eocene Thermal Maximum," *Science Advances* 3(3), e1600891 (2017), https://doi.org/10.1126/sciadv.1600891.

8. Timothy Bralower and David Bice, "Ancient Climate Events: Paleocene Eocene Thermal Maximum," *Earth 103: Earth in the Future*, accessed April 17, 2022, https://www.e-education.psu.edu/earth103/node/639.

9. Paul L. Koch, William C. Clyde, Robert P. Hepple, Marilyn L. Fogel, Scott L. Wing, and James C. Zachos, "Carbon and Oxygen Isotope Records from Paleosols Spanning the Paleocene-Eocene Boundary, Bighorn Basin, Wyoming," in Scott L. Wing, Philip D. Gingerich, Birger Schmitz, and Ellen Thomas (eds.), *Causes and Consequences of Globally Warm Climates in the Early Paleogene*, Special Paper 369 (Boulder, CO: Geological Society of America, 2003), 49–64. Such is the description of Penn State research professor Allie Baczynski: "Paleocene-Eocene Thermal Maximum in Bighorn Basin, Wyoming with Allie Baczynski," *Traveling Geologist*, July 6, 2016, http://www.travelinggeologist.com /2016/07/paleocene-eocene-thermal-maximum-in-bighorn-basin-wyoming -with-allie-baczynski/.

10. Michael Marshall, "Tropical Forests Thrived in Ancient Global Warming," *New Scientist*, November 11, 2010, https://www.newscientist.com/article/ dn19713-tropical-forests-thrived-in-ancient-global-warming/; Gabriel J. Bowen, David J. Beerling, Paul L. Koch, James C. Zachos, and Thomas Quattlebaum, "A Humid Climate State During the Palaeocene/Eocene Thermal Maximum," *Nature* 432, 495–499 (2004), https://doi.org/10.1038/nature03115.

11. See: Robert Kopp, Jonathan Buzan, and Matthew Huber, "The Deadly Combination of Heat and Humidity," *New York Times*, June 6, 2015, https://www .nytimes.com/2015/06/07/opinion/sunday/the-deadly-combination-of-heat -and-humidity.html.

12. See this discussion between experts: Michelle Jewell, Rob Dunn, Ethan Coffel, Steve Sherwood, and Abigail D'Ambrosia, "Tiny Monkey-Horses, Unbearable Heat and Wet Bulbs, an Interview About the Future," *Applied Ecology News*, February 1, 2021, https://cals.ncsu.edu/applied-ecology/news/tiny-monkey -horses-unbearable-heat-and-wet-bulbs-an-interview-about-the-future/; see: Jewell et al., "Tiny Monkey-Horses, Unbearable Heat and Wet Bulbs, an Interview About the Future"; Kim Stanley Robinson, *The Ministry for the Future* (New York: Orbit, 2020); "A Climate Change Q&A with Kim Stanley Robinson & Michael E. Mann" (virtual Q and A), *Orbit LIVE!*, October 7, 2020, https:// www.crowdcast.io/e/climatechangeqa-oct2020/register; Lydia Millet, Kim Stanley Robinson, and Michael Mann, "The Future Is Now for Climate Change," panel at Tucson Festival of Books, March 6, 2021, https://tucsonfestivalofbooks.org/? action=display_event&id=7579&year=2021; Rounak Jain, "Heatwave in India: What Is Wet Bulb Temperature, Why Is It Important and How to Measure It," *Business Insider India*, May 4, 2022, https://www.businessinsider.in/science /environment/news/heatwave-in-india-what-is-wet-bulb-temperature-why-is-it -important-and-how-to-measure-it/articleshow/91306156.cms.

13. Katarzyna B. Tokarska, Nathan P. Gillett, Andrew J. Weaver, Vivek K. Arora, and Michael Eby, "The Climate Response to Five Trillion Tonnes of Carbon," *Nature Climate Change* 6, 851–855 (2016), https://doi.org/10.1038 /nclimate3036; Colin Raymond, Tom Matthews, and Radley M. Horton, "The Emergence of Heat and Humidity Too Severe for Human Tolerance," *Science Advances* 6, eaaw1838 (2020), https://doi.org/10.1126/sciadv.aaw1838; Steven

C. Sherwood and Michael Huber, "An Adaptability Limit to Climate Change Due to Heat Stress," *Science* 107, 9552–9555 (2010), https://doi.org/10.1073/pnas.0913352107.

14. Sourav Mukherjee, Ashok Kumar Mishra, Michael E. Mann, and Colin Raymond, "Anthropogenic Warming and Population Growth May Double US Heat Stress by the Late 21st Century," *Earth's Future* 9(5), e2020EF001886 (2021), https://doi.org/10.1029/2020EF001886; Michael E. Mann, Stefan Rahmstorf, Kai Kornhuber, Byron A. Steinman, Sonya K. Miller, et al., "Projected Changes in Persistent Extreme Summer Weather Events: The Role of Quasi-resonant Amplification," *Science Advances* 4(10), eaat3272 (2018), https://doi.org/10.1126/sciadv.aat3272; Denise Mann, "Workers in U.S. Southwest in Peril as Summer Temperatures Rise," *U.S. News & World Report*, May 18, 2022, https://www.usnews.com/news/health-news/articles/2022-05-18/workers-in-u-s-southwest-in-peril-as-summer-temperatures-rise.

15. Riley Black, "Hot Fossil Mammals May Give a Glimpse of Nature's Future," *National Geographic*, August 13, 2015, https://www.nationalgeographic.com/science/article/hot-fossil-mammals-may-give-a-glimpse-of-natures-future.

16. Ross Secord, Jonathan I. Bloch, Stephen G. B. Chester, Doug M. Boyer, Aaron R. Wood, et al., "Evolution of the Earliest Horses Driven by Climate Change in the Paleocene-Eocene Thermal Maximum," *Science* 535, 959–962 (2012), https://doi.org/10.1126/science.1213859. There is also speculation that higher CO_2 levels might have led to less nutritious leaves, which would have also favored smaller herbivores. See: Penn State, "Insects Will Feast, Plants Will Suffer: Ancient Leaves Show Affect of Global Warming," *ScienceDaily*, February 15, 2008, www.sciencedaily.com/releases/2008/02/080211172638.htm. Examples are provided in: Michael E. Mann, *The New Climate War: The Fight to Take Back Our Planet* (New York: PublicAffairs, 2020); R. Daniel Bressler, "The Mortality Cost of Carbon," *Nature Communications* 12, 4467 (2021), https://doi.org/10.1038/s41467-021-24487-w.

17. Allison Mills, "Boron Proxies Detail Past Ocean Acidification," *Earth Magazine*, September 2, 2014, https://www.earthmagazine.org/article/boron-proxies-detail-past-ocean-acidification/.

18. Christian Robert and James P. Kennett, "Paleocene and Eocene Kaolinite Distribution in the South Atlantic and Southern Ocean: Antarctic Climatic and Paleoceanographic Implications," *Marine Geology* 103, 99–110 (1994), https://doi.org/10.1016/0025-3227(92)90010-F; Karen L. Bice and Jochem Marotzke, "Could Changing Ocean Circulation Have Destabilized Methane Hydrate at the Paleocene/Eocene Boundary?" *Paleoceanography and Paleoclimatology* 17(2), 1018 (2002), https://doi.org/10.1029/2001PA000678; Flavia Nunes and Richard D. Norris, "Abrupt Reversal in Ocean Overturning During the Palaeocene/Eocene Warm Period," *Nature* 439, 60–63 (2006), https://doi.org/10.1038/nature04386.

19. Riley Black, "An Ancient Era of Global Warming Could Hint at Our Scorching Future," *Popular Science*, August 16, 2021, https://www.popsci.com

/environment/petm-climate-change/; Dean Scott, "Warmer Oceans Ahead May Bring More Waves of Toxic Red Tide," *Bloomberg Law*, November 6, 2018, https://news.bloomberglaw.com/environment-and-energy/warmer-oceans -ahead-may-bring-more-waves-of-toxic-red-tide.

20. *Land of the Lost dot com* homepage, accessed May 21, 2022, https:// www.landofthelost.com/.

21. Adam Frank, "Was There a Civilization on Earth Before Humans?" *Atlantic*, April 13, 2018, https://www.theatlantic.com/science/archive/2018/04 /are-we-earths-only-civilization/557180/; Gavin A. Schmidt and Adam Frank, "The Silurian Hypothesis: Would It Be Possible to Detect an Industrial Civilization in the Geological Record?" *International Journal of Astrobiology* 18(2), 142–150 (2018), https://doi.org/10.1017/S1473550418000095.

22. Adam Frank, "A New Frontier Is Opening in the Search for Extraterrestrial Life," *Washington Post*, December 31, 2020, https://www.washington post.com/outlook/2020/12/31/breakthrough-listen-seti-technosignatures/; "Our Founders," *The Planetary Society*, accessed May 22, 2022, https://www .planetary.org/about/our-founders.

23. Keay Davidson, *Carl Sagan: A Life* (New York: John Wiley & Sons, 1999). Sagan friend and protégé David Grinspoon—who is a personal friend of mine—makes a compelling argument in his inspiring book *Earth in Human Hands* (New York: Grand Central Publishing, 2016).

24. Frank, "Was There a Civilization on Earth Before Humans?"

25. Gavin Schmidt, "Under the Sun," *Vice*, April 16, 2018, https://www.vice .com/en/article/3kj4y8/gavin-schmidt-fiction-under-the-sun.

26. For a nice overview, see: Andrea Thompson, "How Did Iceland Form?" *Live Science*, March 22, 2010, https://www.livescience.com/8129-iceland-form .html.

27. A. D. Saunders, S. M. Jones, L. A. Morgan, K. L. Pierce, M. Widdowson, and Y. G. Xu, "Regional Uplift Associated with Continental Large Igneous Provinces: The Roles of Mantle Plumes and the Lithosphere," *Chemical Geology* 241, 282–318 (2007), https://doi.org/10.1016/j.chemgeo.2007.01.017; Michael Storey, Robert A. Duncan, and Carl C. Swisher, III, "Paleocene–Eocene Thermal Maximum and the Opening of the Northeast Atlantic," *Science* 316, 587–589 (2007), https://doi.org/10.1126/science.1135274; Marcus Gutjahr, Andy Ridgwell, Philip F. Sexton, Eleni Anagnostou, Paul N. Pearson, et al., "Very Large Release of Mostly Volcanic Carbon During the Palaeocene–Eocene Thermal Maximum," *Nature* 548, 573–577 (2017), https://doi.org/10.1038 /nature23646; Katrin J. Meissner and Timothy J. Bralower, "Volcanism Caused Ancient Global Warming," *Nature* 548, 531–533 (2017), https://doi .org/10.1038/548531a; Laura L. Haynes and Bärbel Hönisch, "The Seawater Carbon Inventory at the Paleocene–Eocene Thermal Maximum," *Proceedings of the National Academy of Sciences* 117, 24088–24095 (2020), https://doi .org/10.1073/pnas.2003197117.

28. Meissner and Bralower, "Volcanism Caused Ancient Global Warming"; Ryan L. Sriver, Axel Timmermann, Michael E. Mann, Klaus Keller, and Hugues Goosse, "Improved Representation of Tropical Pacific Ocean–Atmosphere Dynamics in an Intermediate Complexity Climate Model," *Journal of Climate* 27(1), 168–187 (2014), https://doi.org/10.1175/JCLI-D-12-00849.1; Hugues Goosse, Joel Guiot, Michael E. Mann, Svetlana Dubinkina, and Yoann Sallaz-Damaz, "The Medieval Climate Anomaly in Europe: Comparison of the Summer and Annual Mean Signals in Two Reconstructions and in Simulations with Data Assimilation," *Global and Planetary Change* 84–85, 35–47 (2012), https://doi.org/10.1016/j.gloplacha.2011.07.002; H. Goosse, E. Crespin, A. de Montety, M. E. Mann, H. Renssen, and A. Timmermann, "Reconstructing Surface Temperature Changes over the Past 600 Years Using Climate Model Simulations with Data Assimilation," *Journal of Geophysical Research* 115, D09108 (2010), https://doi.org/10.1029/2009JD012737; E. Crespin, H. Goosse, T. Fichefet, and M. E. Mann, "The 15th Century Arctic Warming in Coupled Model Simulations with Data Assimilation," *Climate of the Past* 5, 389–405 (2009), https://doi.org/10.5194/cp-5-389-2009.

29. Meissner and Bralower, "Volcanism Caused Ancient Global Warming." A more recent 2020 study based on the analysis of boron and calcium ratios in ocean sediments—a proxy for dissolved inorganic ocean carbon—places the number a bit higher, at just under 14,900 GtC, concluding that it was no greater than eight percent: Haynes and Hönisch, "The Seawater Carbon Inventory at the Paleocene–Eocene Thermal Maximum."

30. Tapio Schneider, Colleen M. Kaul, and Kyle G. Pressel, "Possible Climate Transitions from Breakup of Stratocumulus Decks Under Greenhouse Warming," *Nature Geoscience* 12, 163–167 (2019), https://doi.org/10.1038/s41561-019-0310-1.

31. Paul Voosen, "A World Without Clouds? Hardly Clear, Climate Scientists Say," *Science*, February 26, 2019, https://www.science.org/content/article/world-without-clouds-hardly-clear-climate-scientists-say.

32. Gutjahr et al., "Very Large Release of Mostly Volcanic Carbon During the Palaeocene–Eocene Thermal Maximum." See also: Michael E. Mann, "Beyond the Hockey Stick: Climate Lessons from the Common Era," *Proceedings of the National Academy of Sciences* 118(39), e2112797118 (2021), https://doi.org/10.1073/pnas.2112797118.

33. Rodrigo Caballero and Matthew Huber, "State-Dependent Climate Sensitivity in Past Warm Climates and Its Implications for Future Climate Projections," *Proceedings of the National Academy of Sciences* 110, 14162–14167 (2013), https://doi.org/10.1073/pnas.130336511; Gary Shaffer, Matthew Huber, Roberto Rondanelli, and Jens Olaf Pepke Pedersen, "Deep Time Evidence for Climate Sensitivity Increase with Warming," *Geophysical Research Letters* 43, 6538–6545 (2016), https://doi.org/10.1002/2016GL069243; Gordon N. Inglis, Fran Bragg, Natalie J. Burls, Margot J. Cramwinckel, David Evans, et al.,

"Global Mean Surface Temperature and Climate Sensitivity of the Early Eocene Climatic Optimum (EECO), Paleocene–Eocene Thermal Maximum (PETM), and Latest Paleocene," *Climate of the Past* 16(5), 1953–1968 (2020), https://doi .org/10.5194/cp-16-1953-2020; Tobias Friedrich, Axel Timmermann, Michelle Tigchelaar, Oliver Elison Timm, and Andrey Ganopolski, "Nonlinear Climate Sensitivity and Its Implications for Future Greenhouse Warming," *Science Advances* 2(11), e1501923 (2016), https://doi.org/10.1126/sciadv.1501923; Ivan Mitevski, Lorenzo Polvani, and Clara Orbe, "Asymmetric Warming/Cooling Response to CO_2 Increase/Decrease Mainly Due to Non-logarithmic Forcing, Not Feedbacks," *Geophysical Research Letters* 49, e2021GL097133 (2022), https:// doi.org/10.1029/2021GL097133.

34. Mann, *The New Climate War*.

35. Madeleine Stone, "Earth Is Setting Heat Records. It Will Be Much Hotter One Day," *National Geographic*, August 20, 2020, https://www.nationalgeo graphic.com/science/article/earth-130-degrees-this-week-much-hotter-one-day.

36. Alicia Newton, "Arctic Ice Across the Ages," *Nature Geoscience* 3, 304 (2010), https://doi.org/10.1038/ngeo861; Carolyn D. Ruppel, "Methane Hydrates and Contemporary Climate Change," *Nature Education Knowledge* 3(10), 29 (2011), https://www.nature.com/scitable/knowledge/library/methane -hydrates-and-contemporary-climate-change-24314790/.

37. Timothy Gardner, "Global Methane Emissions Rising Due to Oil and Gas, Agriculture—Studies," *Reuters*, July 14, 2020, https://www.reuters.com /article/us-climate-change-methane/global-methane-emissions-rising-due-to-oil -and-gas-agriculture-studies-idUSKCN24F2X8; Gabrielle B. Dreyfus, Yangyang Xu, Drew T. Shindell, Durwood Zaelke, and Veerabhadran Ramanathan, "Mitigating Climate Disruption in Time: A Self-Consistent Approach for Avoiding Both Near-Term and Long-Term Global Warming," *Proceedings of the National Academy of Sciences* 119(22), e2123536119 (2022), https://doi.org/10.1073 /pnas.2123536119.

38. Michael E. Mann and Susan Joy Hassol, "That Heat Dome? Yeah, It's Climate Change," *New York Times*, June 29, 2021, https://www.nytimes.com /2021/06/29/opinion/heat-dome-climate-change.html.

39. Christopher R. Schwalm, Spencer Glendon, and Philip B. Duffy, "RCP8.5 Tracks Cumulative CO_2 Emissions," *Proceedings of the National Academy of Sciences* 117, 19656–19657 (2020), https://doi.org/10.1073/pnas.2007117117.

40. Valérie Masson-Delmotte, Panmao Zhai, Anna Pirani, Sarah L. Connors, Clotilde Péan, et al. (eds.), *Climate Change 2021: The Physical Science Basis*, Working Group I Contribution to the Sixth Assessment Report of the Intergovernmental Panel on Climate Change (Cambridge, UK: Cambridge University Press, 2021), https://doi.org/10.1017/9781009157896.

41. Peter Brannen, "This Is How Your World Could End," *The Guardian*, September 9, 2017, https://www.theguardian.com/environment/2017/sep/09/this-is -how-your-world-could-end-climate-change-global-warming; Steven C. Sherwood

and Matthew Huber, "An Adaptability Limit to Climate Change Due to Heat Stress," *Science* 107, 9552–9555 (2010), https://doi.org/10.1073/pnas .0913352107.

42. Derek M. Norman, "The 1977 Blackout in New York City Happened Exactly 42 Years Ago," *New York Times*, July 14, 2019, https://www.nytimes .com/2019/07/14/nyregion/1977-blackout-photos.html.

Chapter 6: A Message in the Ice

1. For a particularly vivid description of this period, see: Peter Brannen, "This Is How Your World Could End," *Guardian*, September 9, 2017, https:// www.theguardian.com/environment/2017/sep/09/this-is-how-your-world-could -end-climate-change-global-warming.

2. Eivind O. Straume, Aleksi Nummelin, Carmen Gaina, and Kerim H. Nisancioglu, "Climate Transition at the Eocene–Oligocene Influenced by Bathymetric Changes to the Atlantic–Arctic Oceanic Gateways," *Proceedings of the National Academy of Sciences* 119(17), e2115346119 (2022), https://doi.org/10.1073 /pnas.2115346119; Zhonghui Liu, Mark Pagani, David Zinniker, Robert De- Conto, Matthew Huber, et al., "Global Cooling During the Eocene–Oligocene Climate Transition," *Science* 323, 1187–1190 (2009), https://doi.org/10.1126 /science.1166368.

3. See: "Interval 2," *American Museum of Natural History*, accessed April 17, 2022, https://research.amnh.org/paleontology/perissodactyl/environment/inter val2; "Eocene Epoch (54–33 mya)," *PBS*, accessed April 17, 2022, https://www .pbs.org/wgbh/evolution/change/deeptime/eocene.html; Dorien de Vries, Steven Heritage, Matthew R. Borths, Hesham M. Sallam, and Erik R. Seiffert, "Widespread Loss of Mammalian Lineage and Dietary Diversity in the Early Oligocene of Afro-Arabia," *Communications Biology* 4, 1172 (2021), https://doi .org/10.1038/s42003-021-02707-9.

4. Paul N. Pearson, Gavin L. Foster, and Bridget S. Wade, "Atmospheric Carbon Dioxide Through the Eocene–Oligocene Climate Transition," *Nature* 461, 1110–1113 (2009), https://doi.org/10.1038/nature08447.

5. Robert M. DeConto and David Pollard, "Rapid Cenozoic Glaciation of Antarctica Induced by Declining Atmospheric CO_2," *Nature* 421, 245–249 (2003), https://doi.org/10.1038/nature01290. It should be noted that the precise results appear to be model- and study-dependent. See, for example: Aisling M. Dolan, Bas de Boer, Jorge Bernales, Daniel J. Hill, and Alan M. Haywood, "High Climate Model Dependency of Pliocene Antarctic Ice-Sheet Predictions," *Nature Communications 9*, 2799 (2018), https://doi.org/10.1038/s41467-018-05179-4.

6. Eleni Anagnostou, Eleanor H. John, Kirsty M. Edgar, Gavin L. Foster, Andy Ridgwell, et al., "Changing Atmospheric CO_2 Concentration was the Primary Driver of Early Cenozoic Climate," *Nature* 533, 380–384 (2016), https:// doi.org/10.1038/nature17423; M. Huber and R. Caballero, "The Early Eocene

Equable Climate Problem Revisited," *Climate of the Past* 7(2), 603–633 (2011), https://doi.org/10.5194/cp-7-603-2011; Margot J. Cramwinckel, Matthew Huber, Ilja J. Kocken, Claudia Agnini, Peter K. Bijl, et al., "Synchronous Tropical and Polar Temperature Evolution in the Eocene," *Nature* 559, 382–386 (2018), https://doi.org/10.1038/s41586-018-0272-2.

7. David Pollard and Robert M. DeConto, "Hysteresis in Cenozoic Antarctic Ice-Sheet Variations," *Global and Planetary Change* 45, 9–21 (2005), https://doi.org/10.1016/j.gloplacha.2004.09.011.

8. One notable exception is the eastern and central equatorial Pacific Ocean, which exhibited an El Niño–like warming pattern. See: Marci M. Robinson, Harry J. Dowsett, and Mark A. Chandler, "Pliocene Role in Assessing Future Climate Impacts," *Eos* 89(49), 501–502 (2008), https://doi.org/10.1029/2008eo490001.

9. Eric J. Barron, "A Warm, Equable Cretaceous: The Nature of the Problem," *Earth-Science Reviews* 19(4), 305–338 (1983), https://doi.org/10.1016/0012 8252(83)90001-6; Robert L. Korty and Kerry A. Emanuel, "The Dynamic Response of the Winter Stratosphere to an Equable Climate Surface Temperature Gradient," *Journal of Climate* 20, 5213–5228 (2007), https://doi.org/10.1175 /2007JCLI1556.1. My collaborators and I have also done some research into this mechanism: Ryan L. Sriver, Marlos Goes, Michael E. Mann, and Klaus Keller, "Climate Response to Tropical Cyclone–Induced Ocean Mixing in an Earth System Model of Intermediate Complexity," *Journal of Geophysical Research: Oceans* 115, C10042 (2010), https://doi.org/10.1029/2010JC006106.

10. Richard Levy, David Harwood, Fabio Florindo, Francesca Sangiorgi, Robert Tripati, et al., "Antarctic Ice Sheet Sensitivity to Atmospheric CO_2 Variations in the Early to Mid-Miocene," *Proceedings of the National Academy of Sciences* 113(13), 3453–3458 (2016), https://doi.org/10.1073/pnas.1516030113; Aisling M. Dolan, Bas de Boer, Jorge Bernales, Daniel J. Hill, and Alan M. Haywood, "High Climate Model Dependency of Pliocene Antarctic Ice-Sheet Predictions," *Nature Communications* 9, 2799 (2018), https://doi.org/10.1038/s41467-018 -05179-4; David Pollard and Robert M. DeConto, "Modelling West Antarctic Ice Sheet Growth and Collapse Through the Past Five Million Years," *Nature* 458, 329–332 (2009), https://doi.org/10.1038/nature07809.

11. Maureen E. Raymo, Jerry X. Mitrovica, Michael J. O'Leary, Robert M. DeConto, and Paul J. Hearty, "Departures from Eustasy in Pliocene Sea-Level Records," *Nature Geoscience* 4, 328–332 (2011), https://doi.org/10.1038 /ngeo1118. My former Penn State colleague David Pollard indicated to me that he sees this is a more plausible set of contributions but argues that the true sea level rise might have actually been as small as thirty-three feet, consisting of twenty feet from Greenland, ten feet from the marine portions of the WAIS, and perhaps three feet from the marine portions of the EAIS.

12. See, for example: Gloria Dickie, "Climate Tipping Points of Coral Die-Off, Ice Sheet Collapse Closer than Thought," *Reuters*, September 8, 2022, https://www.reuters.com/business/environment/climate-tipping-points-coral -die-off-ice-sheet-collapse-closer-than-thought-2022-09-08/.

13. Daniel J. Lunt, Alan M. Haywood, Gavin A. Schmidt, Ulrich Salzmann, Paul J. Valdes, and Harry J. Dowsett, "Earth System Sensitivity Inferred from Pliocene Modelling and Data," *Nature Geoscience* 3, 60–64 (2010), https://doi.org/10.1038/ngeo706; IPCC, "Summary for Policymakers," in Valérie Masson-Delmotte, Panmao Zhai, Hans-Otto Pörtner, Debra Roberts, Jim Skea, et al. (eds.), *Global Warming of 1.5°C*, An IPCC Special Report, on the impacts of global warming of 1.5°C above pre-industrial levels and related global greenhouse gas emission pathways, in the context of strengthening the global response to the threat of climate change, sustainable development, and efforts to eradicate poverty (Cambridge, UK: Cambridge University Press, 2018), https://doi.org/10.1017/9781009157940.001.

14. See: Lunt et al., "Earth System Sensitivity Inferred from Pliocene Modelling and Data"; Tony E. Wong, Ying Cui, Dana L. Royer, and Klaus Keller, "A Tighter Constraint on Earth-System Sensitivity from Long-Term Temperature and Carbon-Cycle Observations," *Nature Communications* 12, 3173 (2021), https://doi.org/10.1038/s41467-021-23543-9.

15. Alexander Robinson, Reinhard Calov, and Andrey Ganopolski, "Multistability and Critical Thresholds of the Greenland Ice Sheet," *Nature Climate Change* 2, 429–432 (2012), https://doi.org/10.1038/nclimate1449; Daniel J. Lunt, Gavin L. Foster, Alan M. Haywood, and Emma J. Stone, "Late Pliocene Greenland Glaciation Controlled by a Decline in Atmospheric CO_2 Levels," *Nature* 454, 1102–1105 (2008), https://doi.org/10.1038/nature07223.

16. Robinson et al. ("Multistability and Critical Thresholds of the Greenland Ice Sheet") express their thresholds in terms of a summer temperature anomaly rather than a CO_2 level, making it difficult to interpret, and they don't include Earth orbital cycles, which can influence the likelihood of crossing key temperature thresholds. Ridley et al. find hysteresis but their analysis is complicated by transient effects: Jeff Ridley, Jonathan M. Gregory, Philippe Huybrechts, and Jason Lowe, "Thresholds for Irreversible Decline of the Greenland Ice Sheet," *Climate Dynamics* 35, 1049–1057 (2010), https://doi.org/10.1007/s00382-009-0646-0. Lunt et al. ("Earth System Sensitivity Inferred from Pliocene Modelling and Data") only look at the small-to-large ice sheet transition, and do not state what CO_2 levels were used in their simulations.

17. "The Isthmus of Panama and the Ice Ages," *Science* 287, 13 (2000), https://doi.org/10.1126/science.287.5450.13b.

18. John Turner, Phil Anderson, Tom Lachlan-Cope, Steve Colwell, Tony Phillips, et al., "Record Low Surface Air Temperature at Vostok Station, Antarctica," *Journal of Geophysical Research: Atmospheres* 114(D24), D24102 (2009), https://doi.org/10.1029/2009JD012104.

19. Marc Kaufman, "Russians Drill into Previously Untouched Lake Miles Below Antarctic Glacier," *Washington Post*, February 6, 2012, https://www.washingtonpost.com/national/health-science/russians-drill-into-previously-untouched-lake-vostok-below-antarctica/2012/02/06/gIQAGziNuQ_story.html.

20. Milutin Milankovitch, *Théorie Mathématique des Phénomènes Thermiques Produits par la Radiation Solaire* (Paris: Gauthier-Villars, 1920).

21. John Imbrie and John Z. Imbrie, "Modeling the Climatic Response to Orbital Variations," *Science* 207, 943–953 (1980), https://doi.org/10.1126/science.207.4434.943.

22. Neela Banerjee, "Prominent Climate Change Denier Now Admits He Was Wrong," *Christian Science Monitor*, July 30, 2012, https://www.csmonitor.com/Science/2012/0730/Prominent-climate-change-denier-now-admits-he-was-wrong; "Our Team," *Deep Isolation*, accessed June 2, 2022, https://www.deepisolation.com/team/.

23. "Nemesis Star Theory: The Sun's 'Death Star' Companion," *Space.com*, July 20, 2017, https://www.space.com/22538-nemesis-star.html.

24. Walter Munk, Naomi Oreskes, and Richard Muller, "Gordon James Fraser MacDonald: July 30, 1930–May 14, 2002," in *Biographical Memoirs*, Volume 84 (Washington, DC: The National Academies Press, 2004), 225–250, https://nap.nationalacademies.org/read/10992/chapter/13; Richard A. Muller and Gordon J. MacDonald, "Glacial Cycles and Orbital Inclination," *Nature* 377, 107–108 (1995), https://doi.org/10.1038/377107b0.

25. Richard A. Kerr, "Upstart Ice Age Theory Gets Attentive but Chilly Hearing," *Science* 277, 183–184 (1997), https://doi.org/10.1126/science.277.5323.183. See also: Wallace S. Broecker, David L. Thurber, John Goddard, Teh-Lung Ku, R. K. Matthews, and Kenneth J. Mesolella, "Milankovitch Hypothesis Supported by Precise Dating of Coral Reefs and Deep-Sea Sediments," *Science* 159, 297–300 (1968), https://doi.org/10.1126/science.159.3812.297.

26. In 1962, Saltzman published an article in which he modeled the atmospheric phenomenon of thermal convection through a set of nonlinear equations. Solving these equations numerically, he noted that there was some unstable behavior for certain solutions. A year later, in the same journal, Lorenz published the now-famous article that showed that this system of equations exhibits chaotic behavior—something that was known to be possible theoretically, but which hadn't ever been demonstrated for a real-world physical system. Lorenz credited Saltzman in the acknowledgments: "The writer is indebted to Dr. Barry Saltzman for bringing to his attention the existence of nonperiodic solutions of the convection equations": Edward N. Lorenz, "Deterministic Nonperiodic Flow," *Journal of the Atmospheric Sciences* 20, 130–141 (1963). See also: "Necrologies: Barry Saltzman 1931–2000," *Bulletin of the American Meteorological Society* 82(7), 1448–1450 (2001), https://journals.ametsoc.org/downloadpdf/journals/bams/82/7/1520-0477-82_7_1448.pdf; "In Memoriam: Yale Pioneer in the Theory of Weather and Climate, Barry Saltzman," *Yale University News*, February 5, 2001, https://news.yale.edu/2001/02/05/memoriam-yale-pioneer-theory-weather-and-climate-barry-saltzman.

27. For a representative publication, see: B. Saltzman and M. Y. Verbitsky, "Multiple Instabilities and Modes of Glacial Rhythmicity in the Plio-Pleistocene:

A General Theory of Late Cenozoic Climatic Change," *Climate Dynamics* 9(1), 1–15 (1993), https://doi.org/10.1007/BF00208010.

28. M. Willeit, A. Ganopolski, R. Calov, and V. Brovkin, "Mid-Pleistocene Transition in Glacial Cycles Explained by Declining CO_2 and Regolith Removal," *Science Advances* 5(4), eaav7337 (2019), https://doi.org/10.1126/sciadv.aav7337.

29. A. Ganopolski and R. Calov, "The Role of Orbital Forcing, Carbon Dioxide and Regolith in 100 kyr Glacial Cycles," *Climate of the Past* 7(4), 1415–1425 (2011), https://doi.org/10.5194/cp-7-1415-2011.

30. Nicholas P. McKay, Jonathan T. Overpeck, and Bette L. Otto-Bliesner, "The Role of Ocean Thermal Expansion in Last Interglacial Sea Level Rise," *Geophysical Research Letters* 38(14), L14605 (2011), https://doi.org/10.1029/2011GL048280; Robert M. DeConto and David Pollard, "Contribution of Antarctica to Past and Future Sea-Level Rise," *Nature* 531, 591–597 (2016), https://doi.org/10.1038/nature17145.

31. Tim Stephens, "100,000-Year-Old Polar Bear Genome Reveals Ancient Hybridization with Brown Bears," *UC Santa Cruz News*, June 16, 2022, https://news.ucsc.edu/2022/06/polar-bear-bruno.html; Emma Stone and Alex Farnsworth, "The Last Time Earth Was This Hot Hippos Lived in Britain (That's 130,000 Years Ago)," *The Conversation*, January 20, 2016, https://theconversation.com/the-last-time-earth-was-this-hot-hippos-lived-in-britain-thats-130-000-years-ago-53398; Th. van Kolfschoten, "The Eemian Mammal Fauna of Central Europe," *Netherlands Journal of Geosciences* 79 (2–3), 269–281 (2000), https://doi.org/10.1017/S0016774600021752.

32. Nathaelle Bouttes, "Warm Past Climates: Is Our Future in the Past?" *The National Centre for Atmospheric Science*, August 13, 2018, https://web.archive.org/web/20180813004809/https://www.ncas.ac.uk/en/climate-blog/397-warm-past-climates-is-our-future-in-the-past.

33. Eric Post, Richard B. Alley, Torben R. Christensen, Marc Macias-Fauria, Bruce C. Forbes, et al., "The Polar Regions in a 2°C Warmer World," *Science Advances* 5(12), eaaw9883 (2019), https://doi.org/10.1126/sciadv.aaw9883; Ruediger Stein, Kirsten Fahl, Paul Gierz, Frank Niessen, and Gerrit Lohmann, "Arctic Ocean Sea Ice Cover During the Penultimate Glacial and the Last Interglacial," *Nature Communications* 8, 373 (2017), https://doi.org/10.1038/s41467-017-00552-1; Clara Moskowitz, "Polar Bears Evolved Just 150,000 Years Ago," *Live Science*, March 1, 2010, https://www.livescience.com/10956-polar-bears-evolved-150-000-years.html.

34. Matthew B. Osman, Jessica E. Tierney, Jiang Zhu, Robert Tardif, Gregory J. Hakim, et al., "Globally Resolved Surface Temperatures Since the Last Glacial Maximum," *Nature* 599, 239–244 (2021), https://doi.org/10.1038/s41586-021-03984-4.

35. "La Brea Tar Pits and Hancock Park," *La Brea Tar Pits and Museum*, accessed June 7, 2022, https://tarpits.org/experience-tar-pits/la-brea-tar-pits-and-hancock-park.

36. Fen Montaigne, "The Fertile Shore," *Smithsonian Magazine*, January 2020, https://www.smithsonianmag.com/science-nature/how-humans-came-to-ameri cas-180973739/; David J. Meltzer, "Overkill, Glacial History, and the Extinction of North America's Ice Age Megafauna," *Proceedings of the National Academy of Sciences* 117, 28555–28563 (2020), https://doi.org/10.1073/pnas.201503211.

37. Liz Calvario, "'Before the Flood': Leonardo DiCaprio's Climate Change Doc Gets Record 60 Million Views," *IndieWire*, November 16, 2016, https://www. indiewire.com/2016/11/before-the-flood-climate-change-documentary-record -60-million-views-1201747088/; David Adam, "Gore's Climate Film Has Scientific Errors—Judge," *Guardian*, October 11, 2007, https://www.theguardian.com /environment/2007/oct/11/climatechange.

38. "Convenient Untruths" (group post), *RealClimate*, October 15, 2007, https://www.realclimate.org/index.php/archives/2007/10/convenient-untruths/.

39. Nathan Collins, "A Two-Million-Year History of the Temperature of the Earth," *Pacific Standard*, September 27, 2016, https://psmag.com/news/a-two -million-year-history-of-the-temperature-of-the-earth; Anthea Batsakis, "What Two Million Years of Climate History Tells Us About the Future," *Cosmos Magazine*, September 26, 2016, https://cosmosmagazine.com/earth/climate/what -two-million-years-of-climate-history-tells-us-about-the-future/. See also: Carolyn W. Snyder, "Evolution of Global Temperature over the Past Two Million Years," *Nature* 538, 226–228 (2016), https://doi.org/10.1038/nature19798.

40. Gavin A. Schmidt, Jeff Severinghaus, Ayako Abe-Ouchi, Richard B. Alley, Wallace Broecker, et al., "Overestimate of Committed Warming," *Nature* 547, E16–E17 (2017), https://doi.org/10.1038/nature22803; IPCC, "Summary for Policymakers" (2018).

41. Jessica E. Tierney, Jiang Zhu, Jonathan King, Steven B. Malevich, Gregory J. Hakim, and Christopher J. Poulsen, "Glacial Cooling and Climate Sensitivity Revisited," *Nature* 584, 569–573 (2020), https://doi.org/10.1038/s4 1586-020-2617-x.

42. Scott A. Kulp and Benjamin H. Strauss, "New Elevation Data Triple Estimates of Global Vulnerability to Sea-Level Rise and Coastal Flooding," *Nature Communications* 10, 4844 (2019), https://doi.org/10.1038/s41 467-019-12808-z.

43. Frank Pattyn, Catherine Ritz, Edward Hanna, Xylar Asay-Davis, Rob DeConto, et al., "The Greenland and Antarctic Ice Sheets Under 1.5°C Global Warming," *Nature Climate Change* 8, 1053–1061 (2018), https://doi .org/10.1038/s41558-018-0305-8.

44. Jeff Goodell, "The Doomsday Glacier," *Rolling Stone*, May 9, 2017, https://www.rollingstone.com/politics/politics-features/the-doomsday-glacier -113792/. See the review by my *RealClimate* colleague Stefan Rahmstorf of the Potsdam Institute for Climate Impact Research: "Sea Level in the 5th IPCC Report," *RealClimate*, October 15, 2013, https://www.realclimate.org/index.php /archives/2013/10/sea-level-in-the-5th-ipcc-report/. See also: Kulp and Strauss,

"New Elevation Data Triple Estimates of Global Vulnerability to Sea-Level Rise and Coastal Flooding."

45. J. H. Mercer, "West Antarctic Ice Sheet and CO_2 Greenhouse Effect—Threat of Disaster," *Nature* 271, 321–325 (1978), https://doi.org/10.1038/271321a0.

46. E. Rignot, J. Mouginot, M. Morlighem, H. Seroussi, and B. Scheuchl, "Widespread, Rapid Grounding Line Retreat of Pine Island, Thwaites, Smith, and Kohler Glaciers, West Antarctica, from 1992 to 2011," *Geophysical Research Letters* 41, 3502–3509 (2014), https://doi.org/10.1002/2014GL060140; Goodell, "The Doomsday Glacier."

47. DeConto and Pollard, "Contribution of Antarctica to Past and Future Sea-Level Rise."

48. Katie Hunt, "Massive Amount of Water Found Below Antarctica's Ice Sheet for 1st Time," *CNN*, May 5, 2022, https://www.cnn.com/2022/05/05/world/antarctica-hidden-water-climate-scn/index.html; C-Smart Solutions (@C_Smart_Climate), "Does new mapping of a massive amount of groundwater . . . ," *Twitter*, May 7, 2022, https://twitter.com/C_Smart_Climate/status/1522946760922046464; Personal communication (email) with Richard Alley, May 7, 2022; Prof. Michael E. Mann (@MichaelEMann), "My go-to person on this is my colleague Richard Alley . . . ," *Twitter*, May 7, 2022, https://twitter.com/MichaelEMann/status/1522997499287453697.

49. Pattyn et al., "The Greenland and Antarctic Ice Sheets Under 1.5°C Global Warming."

50. Tamsin L. Edwards, Sophie Nowicki, Ben Marzeion, Regine Hock, Heiko Goelzer, et al., "Projected Land Ice Contributions to Twenty-First-Century Sea Level Rise," *Nature* 593, 74–82 (2021), https://doi.org/10.1038/s41586-021-03302-y; Robert M. DeConto, David Pollard, Richard B. Alley, Isabella Velicogna, Edward Gasson, et al., "The Paris Climate Agreement and Future Sea-Level Rise from Antarctica," *Nature* 593, 83–89 (2021), https://doi.org/10.1038/s41586-021-03427-0.

51. Personal communication (email) with Richard Alley, November 30, 2022.

52. Edwards et al., "Projected Land Ice Contributions to Twenty-First-Century Sea Level Rise"; DeConto et al., "The Paris Climate Agreement and Future Sea-Level Rise from Antarctica."

53. DeConto et al., "The Paris Climate Agreement and Future Sea-Level Rise from Antarctica."

Chapter 7: Beyond the Hockey Stick

1. Michael E. Mann, *The Hockey Stick and the Climate Wars: Dispatches from the Front Lines* (New York: Columbia University Press, 2012).

2. The discussion in this chapter is, in part, adapted from a 2021 review article by the author: Michael E. Mann, "Beyond the Hockey Stick: Climate Lessons

from the Common Era," *Proceedings of the National Academy of Sciences*
118(39), e2112797118 (2021), https://doi.org/10.1073/pnas.2112797118.

3. Tammy M. Rittenour, Julie Brigham-Grette, and Michael E. Mann, "El Niño–
Like Climate Teleconnections in New England During the Late Pleistocene," *Science* 288, 1039–1042 (2000), https://doi.org/10.1126/science.288.5468.1039.

4. Eystein Jansen, Jonathan Overpeck, Keith R. Briffa, Jean-Claude Duplessy, Fortunat Joos, et al., "Palaeoclimate," in Susan Solomon, Dahe Qin, Martin Manning, Z. Chen, Melinda Marquis, et al. (eds.), *Climate Change 2007:
The Physical Science Basis*, Contribution of Working Group I to the Fourth
Assessment Report of the Intergovernmental Panel on Climate Change (Cambridge, UK: Cambridge University Press, 2007), 433–497, https://www.ipcc.ch
/report/ar4/wg1/.

5. See: William F. Ruddiman, "The Early Anthropogenic Hypothesis: Challenges and Responses," *Reviews of Geophysics* 45, RG4001 (2007), https://doi
.org/10.1029/2006RG000207.

6. Olive Heffernan, "Why the Hockey Stick Graph Will Always Be Climate
Science's Icon," *New Scientist*, April 23, 2018, https://www.newscientist.com
/article/2167127-why-the-hockey-stick-graph-will-always-be-climate-sciences
-icon/; Mann, *The Hockey Stick and the Climate Wars*.

7. Carl Sagan, "With Science on Our Side," *Washington Post*, January 9, 1994,
https://www.washingtonpost.com/archive/entertainment/books/1994/01/09
/with-science-on-our-side/9e5d2141-9d53-4b4b-aa0f-7a6a0faff845/; Carl Sagan
and Anne Druyan, *The Demon-Haunted World: Science as a Candle in the Dark*
(New York: Ballantine Books, 1997).

8. Michael E. Mann, Raymond S. Bradley, and Malcolm K. Hughes, "Global-Scale Temperature Patterns and Climate Forcing over the Past Six Centuries," *Nature* 392, 779–787 (1998), https://doi.org/10.1038/33859; Michael
E. Mann, Raymond S. Bradley, and Malcolm K. Hughes, "Northern Hemisphere Temperatures During the Past Millennium: Inferences, Uncertainties,
and Limitations," *Geophysical Research Letters* 26, 759–762 (1999), https://
doi.org/10.1029/1999GL900070; Valérie Masson-Delmotte, Michael Schulz,
Ayako Abe-Ouchi, Jürg Beer, Andrey Ganopolski, et al., "Information from Paleoclimate Archives," in Thomas F. Stocker, Dahe Qin, Gian-Kasper Plattner,
Melinda M. B. Tignor, Simon K. Allen, et al. (eds.), *Climate Change 2013: The
Physical Science Basis*, Contribution of Working Group I to the Fifth Assessment Report of the Intergovernmental Panel on Climate Change (Cambridge,
UK: Cambridge University Press, 2013), 383–464; PAGES 2k Consortium, "A
Global Multiproxy Database for Temperature Reconstructions of the Common
Era," *Scientific Data* 4, 170088 (2017), https://doi.org/10.1038/sdata.2017.88.

9. IPCC, "Summary for Policymakers," in Robert T. Watson, Daniel L. Albritton, Terry Barker, Igor A. Bashmakov, Osvaldo Canziani, et al. (eds.), *Climate
Change 2001: Synthesis Report*, Contribution of Working Groups I, II, and III
to the Third Assessment Report of the Intergovernmental Panel on Climate
Change (Cambridge, UK: Cambridge University Press, 2001), 1–34; Michael E.

Mann, "'Widespread and Severe.' The Climate Crisis Is Here, but There's Still Time to Limit the Damage," *Time*, August 9, 2021, https://time.com/6088531 /ipcc-climate-report-hockey-stick-curve/; IPCC, "Summary for Policymakers," in Valérie Masson-Delmotte, Panmao Zhai, Anna Pirani, Sarah L. Connors, Clotilde Péan, et al. (eds.), *Climate Change 2021: The Physical Science Basis*, Working Group I Contribution to the Sixth Assessment Report of the Intergovernmental Panel on Climate Change (Cambridge, UK: Cambridge University Press, 2021), 3–32.

10. IPCC, "Summary for Policymakers" (2021).

11. D. L. Druckenbrod, M. E. Mann, D. W. Stahle, M. K. Cleaveland, M. D. Therrell, and H. H. Shugart, "Late 18th Century Precipitation Reconstructions from James Madison's Montpelier Plantation," *Bulletin of the American Meteorological Society* 84, 57–71 (2003), https://doi.org/10.1175/BAMS-84-1-57.

12. Druckenbrod et al., "Late 18th Century Precipitation Reconstructions from James Madison's Montpelier Plantation."

13. Thomas Jefferson, *The Writings of Thomas Jefferson*, Volume 16 (Washington, DC: Thomas Jefferson Memorial Association, 1905); quoted in Druckenbrod et al., "Late 18th Century Precipitation Reconstructions from James Madison's Montpelier Plantation."

14. Richard Seager, Mark Cane, Naomi Henderson, Dong-Eun Lee, Ryan Abernathey, and Honghai Zhang, "Strengthening Tropical Pacific Zonal Sea Surface Temperature Gradient Consistent with Rising Greenhouse Gases," *Nature Climate Change* 9, 517–522 (2019), https://doi.org/10.1038/s41558-019-0505-x.

15. Amy C. Clement, Richard Seager, Mark A. Cane, and Stephen E. Zebiak, "An Ocean Dynamical Thermostat," *Journal of Climate* 9, 2190–2196 (1996), https://doi.org/10.1175/1520-0442(1996)009<2190:AODT>2.0.CO;2;   Mark A. Cane, Amy C. Clement, Alexey Kaplan, Yochanan Kushnir, Dmitri Pozdnyakov, et al., "Twentieth-Century Sea Surface Temperature Trends," *Science* 275, 957–960 (1997), https://doi.org/10.1126/science.275.5302.957.

16. Ruben van Hooidonk and Matthew Huber, "Equivocal Evidence for a Thermostat and Unusually Low Levels of Coral Bleaching in the Western Pacific Warm Pool," *Geophysical Research Letters* 36, L06705 (2009), https:// doi.org/10.1029/2008GL036288; Ian N. Williams, Raymond T. Pierrehumbert, and Matthew Huber, "Global Warming, Convective Threshold and False Thermostats," *Geophysical Research Letters* 36, L21805 (2009), https://doi .org/10.1029/2009GL039849; Yiyong Luo, Jian Lu, Fukai Liu, and Oluwayemi Garuba, "The Role of Ocean Dynamical Thermostat in Delaying the El Niño–Like Response over the Equatorial Pacific to Climate Warming," *Journal of Climate* 30, 2811–2827 (2017), https://doi.org/10.1175/JCLI-D-16-0454.1; S. Coats and K. B. Karnauskas, "A Role for the Equatorial Undercurrent in the Ocean Dynamical Thermostat," *Journal of Climate* 31, 6245–6261 (2018), https://doi.org/10.1175/JCLI-D-17-0513.1; Ulla K. Heede, Alexey V. Fedorov, and Natalie J. Burls, "Time Scales and Mechanisms for the Tropical Pacific Response to Global Warming: A Tug of War Between the Ocean Thermostat

and Weaker Walker," *Journal of Climate* 33, 6101–6118 (2020), https://doi .org/10.1175/JCLI-D-19-0690.1.

17. Michael E. Mann, Zhihua Zhang, Scott Rutherford, Raymond S. Bradley, Malcolm K. Hughes, et al., "Global Signatures and Dynamical Origins of the Little Ice Age and Medieval Climate Anomaly," *Science* 326, 1256–1260 (2009), https://doi.org/10.1126/science.1177303; Benjamin I. Cook, Jason E. Smerdon, Richard Seager, and Edward R. Cook, "Pan-Continental Droughts in North America over the Last Millennium," *Journal of Climate* 27, 383–397 (2014), https://doi.org/10.1175/JCLI-D-13-00100.1; Byron A. Steinman, Mark B. Abbott, Michael E. Mann, Joseph D. Ortiz, Song Feng, et al., "Ocean-Atmosphere Forcing of Centennial Hydroclimate Variability in the Pacific Northwest," *Geophysical Research Letters* 41, 2553–2560 (2014), https://doi .org/10.1002/2014GL059499; J. Brad Adams, Michael E. Mann, and Caspar M. Ammann, "Proxy Evidence for an El Niño–Like Response to Volcanic Forcing," *Nature* 426, 274–278 (2003), https://doi.org/10.1038/nature02101; Michael E. Mann, Mark A. Cane, Stephen E. Zebiak, and Amy Clement, "Volcanic and Solar Forcing of the Tropical Pacific over the Past 1000 Years," *Journal of Climate* 18, 447–456 (2005), https://doi.org/10.1175/JCLI-3276.1; Evgeniya Predybaylo, Georgiy Stenchikov, Andrew T. Wittenberg, and Sergey Osipov, "El Niño/Southern Oscillation Response to Low-Latitude Volcanic Eruptions Depends on Ocean Pre-conditions and Eruption Timing," *Communications Earth & Environment* 1, 12 (2020), https://doi.org/10.1038/s43247-020-0013-y; Sylvia G. Dee, Kim M. Cobb, Julien Emile-Geay, Toby R. Ault, R. Lawrence Edwards, et al., "No Consistent ENSO Response to Volcanic Forcing over the Last Millennium," *Science* 367, 1477–1481 (2020), https://doi.org/10.1126 /science.aax2000; Benjamin I. Cook, Toby R. Ault, and Jason E. Smerdon, "Unprecedented 21st Century Drought Risk in the American Southwest and Central Plains," *Science Advances* 1, e1400082 (2015), https://doi.org/10.1126 /sciadv.1400082. See also the discussion in: Mann, "Beyond the Hockey Stick."

18. Cook et al., "Unprecedented 21st Century Drought Risk in the American Southwest and Central Plains."

19. Mann et al., "Global Signatures and Dynamical Origins of the Little Ice Age and Medieval Climate Anomaly."

20. R. P. Acosta and M. Huber, "Competing Topographic Mechanisms for the Summer Indo-Asian Monsoon," *Geophysical Research Letters* 47, e2019GL085112 (2020), https://doi.org/10.1029/2019GL085112; Fangxing Fan, Michael E. Mann, Sukyoung Lee, and Jenni L. Evans, "Observed and Modeled Changes in the South Asian Summer Monsoon over the Historical Period," *Journal of Climate* 23, 5193–5205 (2010), https://doi.org/10.1175 /2010JCLI3374.1; Fangxing Fan, Michael. E. Mann, Sukyoung Lee, and Jenni L. Evans, "Future Changes in the South Asian Summer Monsoon: An Analysis of the CMIP3 Multimodel Projections," *Journal of Climate* 25, 3909–3928 (2012), https://doi.org/10.1175/JCLI-D-11-00133.1.

21. Fangxing Fan, Michael. E. Mann, and Caspar M. Ammann, "Understanding Changes in the Asian Summer Monsoon over the Past Millennium: Insights from a Long-Term Coupled Model Simulation," *Journal of Climate* 22, 1736–1748 (2009), https://doi.org/10.1175/2008JCLI2336.1.

22. L. C. Jackson, R. Kahana, T. Graham, M. A. Ringer, T. Woollings, et al., "Global and European Climate Impacts of a Slowdown of the AMOC in a High Resolution GCM," *Climate Dynamics* 45, 3299–3316 (2015), https://doi.org/10.1007/s00382-015-2540-2.

23. Stefan Rahmstorf, Jason E. Box, Georg Feulner, Michael E. Mann, Alexander Robinson, et al., "Exceptional Twentieth-Century Slowdown in Atlantic Ocean Overturning Circulation," *Nature Climate Change* 5, 475–480 (2015), https://doi.org/10.1038/nclimate2554; L. Caesar, G. D. McCarthy, D. J. R. Thornalley, N. Cahill, and S. Rahmstorf, "Current Atlantic Meridional Overturning Circulation Weakest in Last Millennium," *Nature Geoscience* 14, 118–120 (2021), https://doi.org/10.1038/s41561-021-00699-z.

24. Stefan Hofer, Charlotte Lang, Charles Amory, Christoph Kittel, Alison Delhasse, et al., "Greater Greenland Ice Sheet Contribution to Global Sea Level Rise in CMIP6," *Nature Communications* 11, 6289 (2020), https://doi.org/10.1038/s41467-020-20011-8.

25. Eleanor Frajka-Williams, Isabelle J. Ansorge, Johanna Baehr, Harry L. Bryden, Maria Paz Chidichimo, et al., "Atlantic Meridional Overturning Circulation: Observed Transport and Variability," *Frontiers in Marine Science* 6, 260 (2019), https://doi.org/10.3389/fmars.2019.00260.

26. Bryam Orihuela-Pinto, Matthew H. England, and Andréa S. Taschetto, "Interbasin and Interhemispheric Impacts of a Collapsed Atlantic Overturning Circulation," *Nature Climate Change* (2022), https://doi.org/10.1038/s41558-022-01380-y. See also: Nicola Jones, "Rare 'Triple' La Niña Climate Event Looks Likely—What Does the Future Hold?" *Nature*, June 23, 2022, https://www.nature.com/articles/d41586-022-01668-1.

27. Lijing Cheng, John Abraham, Kevin E. Trenberth, John Fasullo, Tim Boyer, et al., "Another Record: Ocean Warming Continues Through 2021 Despite La Niña Conditions," *Advances in Atmospheric Sciences* 39, 373–385 (2022), https://doi.org/10.1007/s00376-022-1461-3. Record "cold blob" temperatures were reported by my *RealClimate* colleague Stefan Rahmstorf: "Q & A About the Gulf Stream System Slowdown and the Atlantic 'Cold Blob,'" *RealClimate*, October 14, 2016, https://www.realclimate.org/index.php/archives/2016/10/q-a-about-the-gulf-stream-system-slowdown-and-the-atlantic-cold-blob/. Record cold temperatures in the eastern and central equatorial Pacific Ocean were reported by NOAA senior scientist Michael McPhaden during a scientific talk he gave on June 8, 2022: Prof. Matt England (@ProfMatt England), "East and central Pacific SST anomalies at their coldest since 1950 . . .," *Twitter*, June 8, 2022, https://twitter.com/ProfMattEngland/status/15343884 15457660928.

28. See: Stefan Rahmstorf, "The IPCC Sea Level Numbers," *RealClimate*, March 27, 2007, https://www.realclimate.org/index.php/archives/2007/03/the-ipcc-sea-level-numbers/.

29. Stefan Rahmstorf, "A Semi-Empirical Approach to Projecting Future Sea-Level Rise," *Science* 315, 368–370 (2007), https://doi.org/10.1126/science.1135456.

30. Andrew C. Kemp, Benjamin P. Horton, Jeffrey P. Donnelly, Michael E. Mann, Martin Vermeer, and Stefan Rahmstorf, "Climate Related Sea-Level Variations over the Past Two Millennia," *Proceedings of the National Academy of Sciences* 108, 11017–11022 (2011), https://doi.org/10.1073/pnas.1015619108.

31. Robert E. Kopp, Andrew C. Kemp, Klaus Bittermann, Benjamin P. Horton, Jeffrey P. Donnelly, et al., "Temperature-Driven Global Sea-Level Variability in the Common Era," *Proceedings of the National Academy of Sciences* 113, E1434–E1441 (2016), https://doi.org/10.1073/pnas.1517056113.

32. See: Stefan Rahmstorf, "Sea Level in the 5th IPCC Report," *RealClimate*, October 15, 2013, https://www.realclimate.org/index.php/archives/2013/10/sea-level-in-the-5th-ipcc-report/; Stefan Rahmstorf, "Sea Level in the IPCC 6th Assessment Report (AR6)," *RealClimate*, August 13, 2021, https://www.realclimate.org/index.php/archives/2021/08/sea-level-in-the-ipcc-6th-assessment-report-ar6/.

33. Kerry Emanuel, "Response of Global Tropical Cyclone Activity to Increasing CO_2: Results from Downscaling CMIP6 Models," *Journal of Climate* 34, 57–70 (2020), https://doi.org/10.1175/JCLI-D-20-0367.1.

34. See: Jeff Masters, "Above-Normal Atlantic Hurricane Season Is Most Likely This Year: NOAA," *The Weather Underground*, May 25, 2017, https://www.wunderground.com/cat6/above-normal-atlantic-hurricane-season-most-likely-year-noaa.

35. Michael E. Mann, Jonathan D. Woodruff, Jeffrey P. Donnelly, and Zhihua Zhang, "Atlantic Hurricanes and Climate over the Past 1,500 Years," *Nature* 460, 880–883 (2009), https://doi.org/10.1038/nature08219.

36. Andra J. Reed, Michael E. Mann, Kerry A. Emanuel, Ning Lin, Benjamin P. Horton, et al., "Increased Threat of Tropical Cyclones and Coastal Flooding to New York City During the Anthropogenic Era," *Proceedings of the National Academy of Sciences* 112, 12610–12615 (2015), https://doi.org/10.1073/pnas.1513127112; Nathan Rott, "Climate Change's Impact on Hurricane Sandy Has a Price: $8 Billion," *NPR*, May 18, 2021, https://www.npr.org/2021/05/18/997666304/climate-changes-impact-on-hurricane-sandy-has-a-price-8-billion; Andra J. Garner, Michael E. Mann, Kerry A. Emanuel, Robert E. Kopp, Ning Lin, et al., "Impact of Climate Change on New York City's Coastal Flood Hazard: Increasing Flood Heights from the Preindustrial to 2300 CE," *Proceedings of the National Academy of Sciences* 114, 11861–11866 (2017), https://doi.org/10.1073/pnas.1703568114.

37. T. Delworth, S. Manabe, and R. J. Stouffer, "Interdecadal Variations of the Thermohaline Circulation in a Coupled Ocean-Atmosphere Model,"

Journal of Climate 6, 1993–2011 (1993), https://doi.org/10.1175/1520-0442 (1993)006<1993:IVOTTC>2.0.CO;2; Michael E. Mann, Jeffrey Park, and R. S. Bradley, "Global Interdecadal and Century-Scale Climate Oscillations During the Past Five Centuries," *Nature* 378, 266–270 (1995), https://doi.org /10.1038/378266a0.

38. T. L. Delworth and M. E. Mann, "Observed and Simulated Multidecadal Variability in the Northern Hemisphere," *Climate Dynamics* 16, 661–676 (2000), https://doi.org/10.1007/s003820000075. Science writer Richard Kerr wrote a piece about our article for *Science*. In an interview, he asked me what we should call it: Richard A. Kerr, "A North Atlantic Climate Pacemaker for the Centuries," *Science* 288, 1984–1985 (2000), https://doi.org/10.1126/science .288.5473.1984.

39. See the discussion in: Mann, *The Hockey Stick and the Climate Wars*. A review is provided in: Michael E. Mann, Byron A. Steinman, and Sonya K. Miller, "Absence of Internal Multidecadal and Interdecadal Oscillations in Climate Model Simulations," *Nature Communications* 11, 49 (2020), https://doi .org/10.1038/s41467-019-13823-w.

40. Mann et al., "Absence of Internal Multidecadal and Interdecadal Oscillations in Climate Model Simulations."

41. Amanda J. Waite, Jeremy M. Klavans, Amy C. Clement, Lisa N. Murphy, Volker Liebetrau, et al., "Observational and Model Evidence for an Important Role for Volcanic Forcing Driving Atlantic Multidecadal Variability over the Last 600 Years," *Geophysical Research Letters* 47, e2020GL089428 (2020), https://doi.org/10.1029/2020GL089428.

42. The search was performed on June 15, 2022, and yielded the following hits: Addrew Shawn, "Hurricane Season 2022: How Long It Lasts and What to Expect," *Verve Times*, June 8, 2022, https://vervetimes.com/hurricane-season-2022-how -long-it-lasts-and-what-to-expect/; Ryan Smith, "Acrisure Predicts Above-Average Hurricane Season for 2022," *Insurance Business America*, June 2, 2022, https://www.insurancebusinessmag.com/us/news/catastrophe/acrisure -predicts-aboveaverage-hurricane-season-for-2022-408198.aspx; Jairo Ibarra, "Acrisure: 2022 Atlantic Hurricane Season Below 2020, 2021 Activity," *Insurance Insider*, June 1, 2022, https://www.insuranceinsider.com/article/2a69ya394p mcmwe97s16o/catastrophes-section/acrisure-2022-atlantic-hurricane-season -below-2020-2021-activity; "Acrisure Re Issues Qualified Storm Season Prediction," *The Royal Gazette*, June 1, 2022, https://www.royalgazette.com/re-insur ance/business/article/20220601/acrisure-re-issues-qualified-storm-season -prediction/; Karen Braun, "La Nina May Further Disrupt Commodity Markets via Hurricanes," *Reuters*, June 8, 2022, https://www.reuters.com/markets /commodities/la-nina-may-further-disrupt-commodity-markets-via-hurricanes -2022-06-08/.

43. Gabriele C. Hegerl, Thomas J. Crowley, William T. Hyde, and David J. Frame, "Climate Sensitivity Constrained by Temperature Reconstructions over the Past Seven Centuries," *Nature* 440, 1029–1032 (2006), https://doi

.org/10.1038/nature04679; Rick Weiss, "Climate Change Will Be Significant but Not Extreme, Study Predicts," *Washington Post*, April 20, 2006, https://www.washingtonpost.com/wp-dyn/content/article/2006/04/19/AR200604 1902335.html.

44. Michael E. Mann, Jose D. Fuentes, and Scott Rutherford, "Underestimation of Volcanic Cooling in Tree-Ring-Based Reconstructions of Hemispheric Temperatures," *Nature Geoscience* 5, 202–205 (2012), https://doi.org/10.1038/ngeo1394; Michael E. Mann, Scott Rutherford, Andrew Schurer, Simon F. B. Tett, and Jose D. Fuentes, "Discrepancies Between the Modeled and Proxy-Reconstructed Response to Volcanic Forcing over the Past Millennium: Implications and Possible Mechanisms," *Journal of Geophysical Research: Atmospheres* 118, 7617–7627 (2013), https://doi.org/10.1002/jgrd.50609.

45. Scott Rutherford and Michael E. Mann, "Missing Tree Rings and the AD 774–775 Radiocarbon Event," *Nature Climate Change* 4, 648–649 (2014), https://doi.org/10.1038/nclimate2315; Hegerl et al., "Climate Sensitivity Constrained by Temperature Reconstructions over the Past Seven Centuries"; Andrew P. Schurer, Gabriele C. Hegerl, Michael E. Mann, Simon F. B. Tett, and Steven J. Phipps, "Separating Forced from Chaotic Climate Variability over the Past Millennium," *Journal of Climate* 26, 6954–6973 (2013), https://doi.org/10.1175/JCLI-D-12-00826.1.

46. S. C. Sherwood, M. J. Webb, J. D. Annan, K. C. Armour, P. M. Forster, et al., "An assessment of Earth's climate sensitivity using multiple lines of evidence," *Reviews of Geophysics* 58, e2019RG000678 (2020), https://doi.org/10.1029/2019RG000678.

47. United Nations, "Paris Agreement," *United Nations Framework Convention on Climate Change*, 2015, https://unfccc.int/sites/default/files/english_paris_agreement.pdf, Article 2. IPCC, "Summary for Policymakers," in Valérie Masson-Delmotte, Panmao Zhai, Hans-Otto Pörtner, Debra Roberts, Jim Skea, et al. (eds.), *Global Warming of 1.5°C*, An IPCC Special Report, on the impacts of global warming of 1.5°C above pre-industrial levels and related global greenhouse gas emission pathways, in the context of strengthening the global response to the threat of climate change, sustainable development, and efforts to eradicate poverty (Cambridge, UK: Cambridge University Press, 2018), https://doi.org/10.1017/9781009157940.001.

48. Richard J. Millar, Jan S. Fuglestvedt, Pierre Friedlingstein, Joeri Rogelj, Michael J. Grubb, et al., "Emission Budgets and Pathways Consistent with Limiting Warming to 1.5°C," *Nature Geoscience* 10, 741–747 (2017), https://doi.org/10.1038/ngeo3031.

49. A. P. Schurer et al., "Importance of the Pre-industrial Baseline for Likelihood of Exceeding Paris Goals"; A. P. Schurer, K. Cowtan, E. Hawkins, M. E. Mann, V. Scott, and S. F. B. Tett, "Interpretations of the Paris Climate Target," *Nature Geoscience* 11, 220–221 (2018), https://doi.org/10.1038/s41561-018-0086-8.

50. PAGES 2k Consortium, "Consistent Multidecadal Variability in Global Temperature Reconstructions and Simulations over the Common Era,"

Nature Geoscience 12, 643–649 (2019), https://doi.org/10.1038/s41561-019 -0400-0.

Chapter 8: Past Is Prologue. Or Is It?

1. Stephen Schneider spoke at a lecture on November 3, 2009, at the Commonwealth Club in San Francisco, part of which is archived here: Stanford Woods Institute for the Environment, "Stephen Schneider | Climate One Montage" (video), *YouTube*, March 29, 2013, https://www.youtube.com/watch?v =7YZ84pD895Q.

2. Ryan Smith, "Leonardo DiCaprio a 'Sweet Guy,' Says Scientist Who Inspired 'Don't Look Up' Role," *Newsweek*, December 31, 2021, https://www .newsweek.com/leonardo-dicaprio-sweet-guy-michael-mann-scientist-dont-look -role-1664590.

3. See, for example: Rob Waugh, "Warning That Alpine Permafrost 'May Accelerate Global Warming,'" *Yahoo! News*, March 15, 2022, https://news .yahoo.com/warning-that-alpine-permafrost-may-accelerate-global-warming -144856632.html. This story is based on a University of Arizona press release: Mikayla Mace Kelley, "Fast-Melting Alpine Permafrost May Contribute to Rising Global Temperatures," *University of Arizona News*, March 14, 2022, https:// news.arizona.edu/story/fast-melting-alpine-permafrost-may-contribute-rising -global-temperatures.

4. Robert M. DeConto and David Pollard, "Contribution of Antarctica to Past and Future Sea-Level Rise," *Nature* 531, 591–597 (2016), https://doi.org /10.1038/nature17145.

5. Eric Post, Richard B. Alley, Torben R. Christensen, Marc Macias-Fauria, Bruce C. Forbes, et al., "The Polar Regions in a 2°C Warmer World," *Science Advances* 5, eaaw9883 (2019), https://doi.org/10.1126/sciadv.aaw9883.

6. Ruediger Stein, Kirsten Fahl, Paul Gierz, Frank Niessen, and Gerrit Lohmann, "Arctic Ocean Sea Ice Cover During the Penultimate Glacial and the Last Interglacial," *Nature Communications* 8, 373 (2017), https://doi .org/10.1038/s41467-017-00552-1; Clara Moskowitz, "Polar Bears Evolved Just 150,000 Years Ago," *Live Science*, March 1, 2010, https://www.livescience .com/10956-polar-bears-evolved-150-000-years.html.

7. See the thread of comments after this post: Prof. Michael E. Mann (@MichaelEMann), "Another reason that claims that we can rule out the upper end of the climate sensitivity uncertainty range are premature," *Twitter*, October 26, 2020, https://twitter.com/climatedynamics/status/1321068288 697344000.

8. Laura Millan Lombrana, "Climate Change Linked to 5 Million Deaths a Year, New Study Shows," *Bloomberg*, July 7, 2021, https://www.bloomberg .com/news/articles/2021-07-07/climate-change-linked-to-5-million-deaths -a-year-new-study-shows#xj4y7vzkg; "Fossil Fuels May Be Responsible for Twice as Many Deaths as First Thought," *Economist*, February 25, 2021, https://

www.economist.com/graphic-detail/2021/02/25/fossil-fuels-may-be-responsible
-for-twice-as-many-deaths-as-first-thought.

9. Denise Mann, "Workers in U.S. Southwest in Peril as Summer Temperatures
Rise," *U.S. News & World Report*, May 18, 2022, https://www.usnews.com
/news/health-news/articles/2022-05-18/workers-in-u-s-southwest-in-peril-as
-summer-temperatures-rise; Zachary Hansen, "It's So Hot in Phoenix, They Can't
Fly Planes," *AZ Central*, June 20, 2017, https://www.azcentral.com/story/travel
/nation-now/2017/06/19/its-so-hot-phoenix-they-cant-fly-planes/410766001/.

10. "Maricopa's Ozone High Pollution Advisory Extended Through Tuesday,
June 20, 2017," *Phoenix Interagency Dispatch Center*, June 19, 2017, https://
www.az-phc.com/2017/06/page/36/.

11. Michael Mann, "Australia, Your Country Is Burning—Dangerous Climate Change Is Here with You Now," *Guardian*, January 1, 2020, https://www
.theguardian.com/commentisfree/2020/jan/02/australia-your-country-is-burning
-dangerous-climate-change-is-here-with-you-now.

12. "'Unprecedented' South Asian Heatwave 'Testing the Limits of Human
Survivability,'" *Climate Signals*, May 3, 2022, https://www.climatesignals.org
/headlines/unprecedented-south-asian-heatwave-testing-limits-human-survivability.

13. "Western North American Extreme Heat Virtually Impossible Without
Human-Caused Climate Change," *World Weather Attribution*, July 7, 2021,
https://www.worldweatherattribution.org/western-north-american-extreme
-heat-virtually-impossible-without-human-caused-climate-change/.

14. Michael E. Mann and Susan Joy Hassol, "That Heat Dome? Yeah, It's
Climate Change," *New York Times*, June 29, 2021, https://www.nytimes.com
/2021/06/29/opinion/heat-dome-climate-change.html.

15. Michael E. Mann, Stefan Rahmstorf, Kai Kornhuber, Byron A. Stein-
man, Sonya K. Miller, and Dim Coumou, "Influence of Anthropogenic Climate
Change on Planetary Wave Resonance and Extreme Weather Events," *Scientific
Reports* 7, 45242 (2017), https://doi.org/10.1038/srep45242; Michael E. Mann,
Stefan Rahmstorf, Kai Kornhuber, Byron A. Steinman, Sonya K. Miller, et al.,
"Projected Changes in Persistent Extreme Summer Weather Events: The Role of
Quasi-resonant Amplification," *Science Advances* 4, eaat3272 (2018), https://
doi.org/10.1126/sciadv.aat3272.

16. My colleague Kai Kornhuber of Columbia University demonstrated
that resonance conditions held at the time: Prof. Michael E. Mann (@Michael
EMann), "The extreme weather we're seeing in the Northern Hemisphere appears to be a consequence of 'resonance' . . . ," *Twitter*, June 20, 2022, https://
twitter.com/MichaelEMann/status/1538892180521115649; Susan Joy Hassol and Michael E. Mann, "Heat Wave Bakes One-Third of Americans, Highlighting Urgency of Climate Legislation," *The Hill*, June 15, 2022, https://
thehill.com/opinion/energy-environment/3524705-heat-wave-bakes-one-third
-of-americans-highlighting-urgency-of-climate-legislation; Gabrielle Canon,
"'Historic' Weather: Why a Cocktail of Natural Disasters Is Battering the
US," *Guardian*, June 18, 2022, https://www.theguardian.com/us-news/2022

/jun/17/compound-extremes-natural-disasters-us-west; USA TODAY Network and the Associated Press, "Here Are All the People Who Died in the California Mudslide," *USA TODAY*, January 14, 2018, https://www.usatoday.com /story/news/nation-now/2018/01/13/all-people-who-died-california-mudslides /1031202001/; Gabrielle Canon, "California Storm Death Toll Climbs to 20 as Deluge Begins to Subside," *Guardian*, January 17, 2023, https://www.theguardian .com/us-news/2023/jan/17/california-storm-death-toll-deluge-subside.

17. Julhas Alam and Wasbir Hussain, "Millions of Homes Under Water as Floods Hit India, Bangladesh," *Sydney Morning Herald*, June 18, 2022, https://www.smh.com.au/world/asia/millions-of-homes-under-water-as-floods -hit-india-bangladesh-20220618-p5auqr.html; Paolo Santalucia, "As Po Dries Up, Italy's Food and Energy Supplies Are at Risk," *Associated Press*, June 17, 2022, https://apnews.com/article/climate-italy-and-environment-e0274e5f2b4 dd6bb2854cc7a970f75f6; Michael E. Mann, "It's Not Rocket Science: Climate Change Was Behind This Summer's Extreme Weather," *Washington Post*, November 2, 2018, https://www.washingtonpost.com/opinions/its-not-rocket -science-climate-change-was-behind-this-summers-extreme-weather/2018 /11/02/b8852584-dea9-11e8-b3f0-62607289efee_story.html.

18. Mann et al., "Projected Changes in Persistent Extreme Summer Weather Events: The Role of Quasi-resonant Amplification."

19. Warren Cornwall, "Even 50-Year-Old Climate Models Correctly Predicted Global Warming," *Science*, December 4, 2019, https://science.org/content /article/even-50-year-old-climate-models-correctly-predicted-global-warming.

20. Adam Quinton (@adamquinton), "Again we see that climate science as often presented to the public is too conservative . . . ," *Twitter*, June 18, 2022, https://web.archive.org/web/20220618121210/https://twitter.com/adam quinton/status/1538132292228489216.

21. Prof. Michael E. Mann (@MichaelEMann), "Actually, the warming of the planet is very much in line with early climate model predictions . . . ," *Twitter*, June 18, 2022, https://twitter.com/MichaelEMann/status/1538175708471648256; Stanford Woods Institute for the Environment, "Stephen Schneider | Climate One Montage."

22. Jamie Gumbrecht, "Formula Production at Abbott's Michigan Plant Delayed After Flooding from Severe Storms," *CNN*, June 16, 2022, https://www .cnn.com/2022/06/15/health/abbott-formula-plant-flood-delay/index.html; Ciara Nugent, "Rising Heat Is Making It Harder to Work in the U.S., and the Costs to the Economy Will Soar with Climate Change," *Time*, August 31, 2021, https:// time.com/6093845/how-heat-hurts-the-economy/.

23. "UN Report: Nature's Dangerous Decline 'Unprecedented'; Species Extinction Rates 'Accelerating,'" United Nations press release, May 6, 2019, https:// www.un.org/sustainabledevelopment/blog/2019/05/nature-decline-unprece dented-report/; IPCC [Hans-Otto Pörtner, Debra C. Roberts, Helen Adams, Carolina Adler, Paulina Aldunce, et al.], "Summary for Policymakers," in Hans-Otto Pörtner, Debra C. Roberts, Melinda M. B. Tignor, Elvira Poloczanska, Katja

Mintenbeck, et al. (eds.), *Climate Change 2022: Impacts, Adaptation, and Vulnerability*, Working Group II Contribution to the Sixth Assessment Report of the Intergovernmental Panel on Climate Change (Cambridge, UK: Cambridge University Press, 2022).

24. See: Mark Hertsgaard, Saleemul Huq, and Michael E. Mann, "How a Little-Discussed Revision of Climate Science Could Help Avert Doom," *Washington Post*, February 23, 2022, https://www.washingtonpost.com/outlook /2022/02/23/warming-timeline-carbon-budget-climate-science/.

25. Myles R. Allen, Opha Pauline Dube, William Solecki, Fernando Aragón-Durand, Wolfgang Cramer, et al., "Framing and Context," in Valérie Masson-Delmotte, Panmao Zhai, Hans-Otto Pörtner, Debra Roberts, Jim Skea, et al. (eds.), *Global Warming of 1.5°C*, An IPCC Special Report, on the impacts of global warming of 1.5°C above pre-industrial levels and related global greenhouse gas emission pathways, in the context of strengthening the global response to the threat of climate change, sustainable development, and efforts to eradicate poverty (Cambridge, UK: Cambridge University Press, 2018), 49–92, https://doi.org/10.1017/9781009157940.003.

26. Malte Meinshausen, Jared Lewis, Christophe McGlade, Johannes Gütschow, Zebedee Nicholls, et al., "Realization of Paris Agreement Pledges May Limit Warming Just Below 2°C," *Nature* 604, 304–309 (2022), https://doi .org/10.1038/s41586-022-04553-z; Michael E. Mann and Susan Joy Hassol, "Glasgow's Hope at a Critical Moment in the Climate Battle," *Los Angeles Times*, November 13, 2021, https://www.latimes.com/opinion/story/2021-11-13/cop 26-glasgow-climate-change.

27. Damian Carrington and Matthew Taylor, "Revealed: The 'Carbon Bombs' Set to Trigger Catastrophic Climate Breakdown," *Guardian*, May 11, 2022, https://www.theguardian.com/environment/ng-interactive/2022/may/11/fossil -fuel-carbon-bombs-climate-breakdown-oil-gas.

28. See: Anthony Leiserowitz, Edward Maibach, Seth Rosenthal, John Kotcher, Jennifer Carman, et al., "Public Support for Climate Action by the President and Congress Is Rising," *Yale Program on Climate Change Communication*, September 28, 2021, https://climatecommunication.yale.edu/publications /public-support-for-climate-action-by-the-president-and-congress/; Oliver Milman, "Republicans Pledge Allegiance to Fossil Fuels Like It's Still the 1950s," *Guardian*, June 7, 2021, https://www.theguardian.com/us-news/2021/jun/07 /republicans-fossil-fuels-coal.

29. Michael E. Mann, *The New Climate War: The Fight to Take Back Our Planet* (New York: PublicAffairs, 2020); Nick Grimm, "Rupert and the Saudi Prince: Key Murdoch Ally Sells Off Shares in 21st Century Fox," *Australian Broadcast Corporation*, November 7, 2017, https://www.abc.net.au/news/2017-11-08 /key-murdoch-ally-saudi-prince-sells-shares-in-21st-century-fox/9129470.

30. See, for example: Robin Young and Serena McMahon, "'Climate Denial' to 'Climate Delay': Rupert Murdoch's News Corp Pivots Media

Narrative in Australia," *WBUR*, November 9, 2021, https://www.wbur.org/hereandnow/2021/11/09/climate-change-tabloids-murdoch.

31. For a more detailed discussion of these events, see: Mann, *The New Climate War*.

32. Julia Baird, "A Carbon Tax's Ignoble End," *New York Times*, July 24, 2014, https://www.nytimes.com/2014/07/25/opinion/julia-baird-why-tony-abbott-axed-australias-carbon-tax.html.

33. I had the pleasure of meeting with Zali Steggall at her seaside office in Manly in mid-February 2020. Manly is a peninsula that juts out into the Pacific Ocean, forming the northern gate of Sydney Harbor. It's home to one of Australia's prized beaches (my family and I spent New Year's Day 2020 there). I have to confess to having been just a bit intimidated by Steggall going into the meeting. Was it because she's a four-time Olympian whose bronze medal at Nagano in 1998 was only the second Winter Olympic medal in Australia's history? Perhaps. But I think it's mostly because she's just plain badass.

34. Michael Mann and Malcolm Turnbull, "How Australia's Electoral System Allowed Voters to Finally Impose a Ceasefire in the Climate Wars," *Guardian*, May 27, 2022, https://www.theguardian.com/commentisfree/2022/may/28/how-australias-electoral-system-allowed-voters-to-finally-impose-a-ceasefire-in-the-climate-wars.

35. Mann and Turnbull, "How Australia's Electoral System Allowed Voters to Finally Impose a Ceasefire in the Climate Wars."

36. Zing Tsjeng, "The Climate Change Paper So Depressing It's Sending People to Therapy," *Vice*, February 27, 2019, https://www.vice.com/en/article/vbwpdb/the-climate-change-paper-so-depressing-its-sending-people-to-therapy; Jem Bendell, "Deep Adaptation: A Map for Navigating Climate Tragedy," IFLAS Occasional Paper 2, July 27, 2018, https://web.archive.org/web/20180805214026/https://www.lifeworth.com/deepadaptation.pdf; Thomas Nicholas, Galen Hall, and Colleen Schmidt, "The Faulty Science, Doomism, and Flawed Conclusions of Deep Adaptation," *Scientists' Warning*, July 14, 2020, https://www.scientistswarning.org/2020/07/14/the-faulty-science-doomism-and-flawed-conclusions-of-deep-adaptation/; Tsjeng, "The Climate Change Paper So Depressing It's Sending People to Therapy"; Jack Hunter, "The 'Climate Doomers' Preparing for Society to Fall Apart," *BBC News*, March 16, 2020, https://www.bbc.com/news/stories-51857722.

37. Susan Joy Hassol and Michael E. Mann, "Now Is Not the Time to Give in to Climate Fatalism," *Time*, April 12, 2022, https://time.com/6166123/climate-change-fatalism/.

38. Samantha K. Stanley, Teaghan L. Hogg, Zoe Leviston, and Iain Walker, "From Anger to Action: Differential Impacts of Eco-anxiety, Eco-depression, and Eco-anger on Climate Action and Wellbeing," *The Journal of Climate Change and Health* 1, 100003 (2021), https://doi.org/10.1016/j.joclim.2021.100003.

Index

See also specific ice ages
ice albedo feedback, 62–63, 68, 79,
      150, 152
ice cliffs, 176
ice core research, 161
ice elevation feedback, 150
ice sheets
   in Antarctica, 150
   building of, 63–64, 65, 148
   collapse as threat, 174
   in Eocene, 148
   expansion, 13–14
   feedback mechanisms in, 155, 176
   groundwater underneath, 177–178
   and hysteresis, 157–158
   melting (see melting of ice sheets)
   in Pleistocene, 159–160, 173–174
   in Pliocene, 154, 221–222
   and post-glacial rebound, 156
   and regolith, 166–167
   and sea level, 155–156, 174,
      177–179
   sediments from, 183–184
   See also individual ice sheets
icehouse climate, 11
Iceland, 137–139
Imbrie, John, 163
India, wet bulb temperature,
      127–128, 228
industrial civilization, impact on
      climate, 38–39, 185, 215
insect species, 72
interglacial cycles. See glacial/
      interglacial cycles
Intermediate-Range Nuclear Forces
      (INF) Treaty, 119
IPCC (Intergovernmental Panel on
      Climate Change)
   hockey stick curve, 187 (fig.), 188
   scenarios used, 225–226
   sea level rise projections, 201,
      203–204
   targets, 234
   warming baseline, 214

warming increase as scenario, 145
iridium, 94
irrigation, and agriculture, 22–24

Jarmo (village), 21
Jastrow, Robert, 110–111
Jefferson, Thomas, 190, 191–193
Jericho (city), 20
jet streams, 27, 31, 171, 195, 196,
      229
Joughin, Ian, 176

K-Pg boundary, 94, 95, 116–117
K-Pg extinction event
   animals in, 95, 97, 98, 98 (fig.), 120
   cause, 95, 114–115, 118, 219
   denial of, 114
   and dinosaurs' extinction, 10,
      94–95, 115
   oceans, 115–116, 117–118
kaolinite, 131
Keeling, Charles, 161
Kemp, Andrew, 202
Kerr, Richard, 165
Kirschvink, Joseph, 65, 66
Kump, Lee, and colleagues, 61, 85,
      86, 87, 88–89

La Brea Tar Pits Museum, 170–171,
      170 (fig.)
La Niña, 27, 193, 194–195, 205
Land of the Lost (TV series),
      132–133
Last Glacial Maximum (LGM), 14,
      141, 142, 162, 168, 169–174,
      223
Late Antique Little Ice Age (LALIA),
      30
latitude, 63, 154
Laurentide Ice Sheet, 15, 20, 156,
      159–160, 183
Little Ice Age, 2–3, 32, 34, 195–196
Living Ocean model, 116–117
lizard people in 1970s, 132–133

MICHAEL E. MANN is presidential distinguished professor and director of the Center for Science, Sustainability, and the Media at the University of Pennsylvania.

He has received many honors and awards, including NOAA's outstanding publication award in 2002 and selection by *Scientific American* as one of the fifty leading visionaries in science and technology in 2002. Additionally, he contributed, with other IPCC authors, to the award of the 2007 Nobel Peace Prize.

More recently, he received the Award for Public Engagement with Science from the American Association for the Advancement of Science in 2018 and the Climate Communication Prize from the American Geophysical Union in 2018. In 2019 he received the Tyler Prize for Environmental Achievement. In 2020 he was elected to the U.S. National Academy of Sciences. He is the author of numerous books, including *Dire Predictions: Understanding Climate Change*, *The Hockey Stick and the Climate Wars: Dispatches from the Front Lines*, and *The Madhouse Effect: How Climate Change Denial Is Threatening Our Planet, Destroying Our Politics, and Driving Us Crazy*. He splits his time these days between Philadelphia and State College, Pennsylvania.

PublicAffairs is a publishing house founded in 1997. It is a tribute to the standards, values, and flair of three persons who have served as mentors to countless reporters, writers, editors, and book people of all kinds, including me.

I. F. STONE, proprietor of *I. F. Stone's Weekly*, combined a commitment to the First Amendment with entrepreneurial zeal and reporting skill and became one of the great independent journalists in American history. At the age of eighty, Izzy published *The Trial of Socrates*, which was a national bestseller. He wrote the book after he taught himself ancient Greek.

BENJAMIN C. BRADLEE was for nearly thirty years the charismatic editorial leader of *The Washington Post*. It was Ben who gave the *Post* the range and courage to pursue such historic issues as Watergate. He supported his reporters with a tenacity that made them fearless and it is no accident that so many became authors of influential, best-selling books.

ROBERT L. BERNSTEIN, the chief executive of Random House for more than a quarter century, guided one of the nation's premier publishing houses. Bob was personally responsible for many books of political dissent and argument that challenged tyranny around the globe. He is also the founder and longtime chair of Human Rights Watch, one of the most respected human rights organizations in the world.

· · ·

For fifty years, the banner of Public Affairs Press was carried by its owner Morris B. Schnapper, who published Gandhi, Nasser, Toynbee, Truman, and about 1,500 other authors. In 1983, Schnapper was described by *The Washington Post* as "a redoubtable gadfly." His legacy will endure in the books to come.

Peter Osnos, *Founder*